T0212893

Developments in Mathematics

VOLUME 33

Series Editors:

Krishnaswami Alladi, *University of Florida*
Hershel M. Farkas, *Hebrew University of Jerusalem*

For further volumes:
http://www.springer.com/series/5834

Christopher S. Hardin • Alan D. Taylor

The Mathematics of Coordinated Inference

A Study of Generalized Hat Problems

 Springer

Christopher S. Hardin
Jane Street
New York, NY, USA

Alan D. Taylor
Department of Mathematics
Union College
Schenectady, NY, USA

ISSN 1389-2177
ISBN 978-3-319-37605-9   ISBN 978-3-319-01333-6 (eBook)
DOI 10.1007/978-3-319-01333-6
Springer Cham Heidelberg New York Dordrecht London

Mathematics Subject Classification (2010): 03E05, 03E17

Printed on acid-free paper

Springer is part of Springer Science+Business Media (www.springer.com)

To Novem

C.S.H.

To Joel and Alan

A.D.T.

Preface

This book deals with the question of how successfully one can predict the value of an arbitrary function at one or more points of its domain based on some knowledge of its values at other points. In large part because of the axiom of choice, the degree of success turns out to be quite remarkable in a number of different situations. For example, there is a method of predicting the value $f(a)$ of a function f mapping the reals to the reals, based only on knowledge of f's values on the interval $(a - 1, a)$, and for every such function the prediction is incorrect only on a countable set that is nowhere dense. In many cases we are able to show that there is a "predictor" that is as successful as possible, and that it is essentially unique.

We collect together most of what is known regarding this problem and we provide a number of extensions of the results that we have published in a series of seven papers over the past six years. We also indicate a number of open questions, ranging from the combinatorially difficult finite to those arising from some results that we show are independent of ZFC plus a fixed value of the continuum.

New York, NY, USA   Christopher S. Hardin
Schenectady, NY, USA   Alan D. Taylor

Contents

Chapter 1
Introduction

1.1 Background

Hat problems trace their roots at least as far back as Martin Gardner's 1961 article (republished as [Gar61]), and in the infinite case to Fred Galvin's work in the mid 1960s [Gal65]. A surge of interest in hat problems during the past decade or so was largely inspired by some popular press based on a hat problem presented by Todd Ebert in his 1998 PhD thesis [Ebe98]. While Ebert's hat problem is certainly of interest, the following variant is really the inspiration for most of what we do here, and it, interestingly, predates Ebert's version (see [ABFR94]).

Two prisoners are told that they will be brought to a room and seated so that each can see the other. Hats will be placed on their heads and each hat is either red or green. The two prisoners must simultaneously submit a guess of their own hat color, and they both go free if at least one of them guesses correctly. While no communication is allowed once the hats have been placed, they will, however, be allowed to have a strategy session before being brought to the room. Is there a strategy ensuring their release?

This puzzle is instructive in (at least) two different ways. First, the solution— which is essentially unique—can come either by an inexplicable flash of insight or by an exhaustive (but hardly exhausting) examination of the 16 pairs of individual strategies arising from the observation that each prisoner can do only one of four things: he can always say red, he can always say green, he can guess the color he sees, or he can guess the color he doesn't see. Second, the solution, once found, can be phrased in two different ways, one of which requires a moment's thought to see why it works and one of which doesn't. The former is: "One prisoner guesses the hat color he sees and the other guesses the hat color he doesn't see," and the latter is: "One prisoner assumes the hats are the same color while the other prisoner assumes they are not."

But what if there are more than two prisoners and more than two hat colors? And what if each prisoner sees only some of the other hats, and their collective freedom depends on several of them guessing correctly? Not surprisingly, the innocent two-person hat problem above quickly leads to interesting combinatorics, both finitary and infinitary, and an array of open questions, with some natural variants having important connections with extant areas like coding theory, ultrafilters, and cardinal invariants of the continuum.

There is a reason that these hat problems are often stated in the metaphorical setting of prisoners trying to secure their collective release (or players who will share equally in a monetary prize), and this is the issue that led to our use of "coordinated inference" in the title of this monograph. Hendrik Lenstra and Gadiel Seroussi [LS02] speak of a "strategy coordination meeting" before the hats are placed. The individual players or agents ("prisoners") are choosing a strategy as they would in any game, but instead of pitting these strategies against those of the other agents, they are seeking to coordinate these strategy choices into a "meta-strategy" that will be pitted against a single adversary who will be placing the hats.

We shall, however, largely move away from the prisoners metaphor, speaking instead of "agents," as we turn to our main focus in this monograph, the infinite. The kind of strategies we are interested in were called "public deterministic strategies" in [BHKL08]. They are public in the sense that the adversary placing the hats is aware of the agents' strategies and they are deterministic in the sense that guesses are uniquely determined, rather than relying on some element of randomness. This determinism leads to a natural focus on worst-case scenarios as opposed to average-case scenarios.

A reasonably general framework for hat problems is the one in which we have a set A (of *agents*), a set K of (*colors*), and a set C of functions (*colorings*) mapping A to K. The goal is for the agents to construct coordinated strategies so that if each agent is given a certain piece of information about one of the colorings, then he can provide a guess as to some other aspect of the coloring. The collection of guesses, taken together over the set of agents, picks out a (possibly empty) set of colorings—those consistent with every agent's guess. We think of this process of collecting together the guesses of the agents as a "predictor." In most cases of interest, this prediction will be a single coloring. Indeed, when we formalize the notion of predictor below, we do so in a context considerably less general than the natural framework just described.

The information provided to an agent $a \in A$ is typically captured by an equivalence relation $\equiv_a$ on the set of colorings, the intuition being that $f \equiv_a g$ indicates that agent a cannot distinguish between the coloring f and the coloring g. There is also a notion of when a prediction is successful enough to be deemed acceptable. Much (but not all) of our interest is in predictors that are either *minimal* in the sense of achieving the least success in terms of the number of agents guessing correctly that could be called nontrivial, or *optimal* in the (rough) sense of achieving the most success that is possible in the given context.

In many cases of interest, the equivalence relations $\{ \equiv_a : a \in A \}$ will be given by a directed graph V on A that we call the *visibility graph*, with the intended meaning of an edge $a \rightarrow b$ being that agent a can see (the hat worn by) agent b. Formally, V is an irreflexive binary relation on A, with edges $a \rightarrow b$ identified with pairs (a, b). We adopt the common shorthand aVb for $(a, b) \in V$, and let

$V(a) = \{\, b : aVb \,\}$, the set of agents visible to a. Accordingly, we let $f \equiv_a g$ iff $f(b) = g(b)$ for all $b \in V(a)$: a cannot distinguish f and g if they assign the same colors to all hats a can see.

Thus, although we later move to more general contexts, we can for now think of a hat problem as made up of a set A of agents, a set K of colors, a set C of colorings, each of which maps A to K, and a collection of equivalence relations $\{\equiv_a : a \in A\}$ on C. A *guessing strategy* for agent a is a function $G_a : C \to K$ with the property that if $f \equiv_a g$, then $G_a(f) = G_a(g)$. Intuitively, agent a is guessing the color of his hat under the hat coloring f, and this guess must be the same for any two colorings that he cannot distinguish between. We say that *agent a guesses correctly* if $G_a(f) = f(a)$. Given this sequence of guessing strategies, the corresponding *predictor* is then defined to be the map $P : C \to C$ given by $P(f)(a) = G_a(f)$.

A predictor's success is measured, at least in part, by the "size" of the set of agents guessing correctly. In the finite case, size will correspond to cardinality, but in the infinite case there are more refined notions, many of which involve the concept of an ideal, which we present in the next section.

The rest of this chapter is organized as follows. In Sect. 1.2 we collect together the set-theoretic preliminaries that arise in two or more sections of the monograph; concepts like P-point ultrafilters that occur in only one section are introduced where they occur. In Sect. 1.3 we collect together several techniques that will frequently be used to show the non-existence of certain kinds of predictors, and in Sect. 1.4 we present one very successful predictor—the μ-predictor—which will arise repeatedly in different guises. Finally, in Sect. 1.5 we give a brief chapter-by-chapter overview of what is to follow.

1.2 Set-Theoretic Preliminaries

Our set-theoretic notation and terminology are reasonably standard. If A is a set, then $\mathcal{P}(A)$ is the power set of A and $|A|$ is the cardinality of A. We use $[A]^k$ to denote the set of all k-element subsets of A, $[A]^\omega$ to denote the set of all countably infinite subsets of A, and $[A]^{<\lambda}$ to denote the collection of subsets of A of cardinality less than λ. If f is a function, then $f|X$ is the restriction of f to X, and $^A K$ is the set of functions mapping the set A into the set K. If X and Y are sets, then $X - Y = \{x \in X : x \notin Y\}$ and $X \Delta Y = (X - Y) \cup (Y - X)$. If f and g are functions, then $f \Delta g = \{x : f(x) \neq g(x)\}$. For a real number x, we use $\lfloor x \rfloor$ for the largest integer not greater than x and $\lceil x \rceil$ for the smallest integer not less than x.

Each ordinal is identified with the set of smaller ordinals, so ω is the set of natural numbers, $n = \{0, 1, \ldots, n-1\}$, and, in particular, $2 = \{0, 1\}$. ZF denotes the Zermelo-Fraenkel axioms for set theory, ZFC is ZF plus the axiom of choice, and DC is the axiom of dependent choices (the assertion that if R is a binary relation on X such that for every $x \in X$ there exists $y \in X$ such that xRy, then there exists a sequence $\langle x_n : n \in \omega \rangle$ such that $x_n R x_{n+1}$ for every $n \in \omega$).

We will have several occasions to make use of the special case of Ramsey's theorem [Ram30] that asserts that an infinite graph has either an infinite complete subgraph or an infinite independent subgraph. Generalizations of this result (and the standard notation used) will be introduced as they arise.

The topology and measure on $^\omega 2$ are the usual ones. That is, if $X \in [\omega]^n$ and $s : X \to 2$ then the set $[s] = \{ g \in {}^\omega 2 : g|X = s \}$ is a basic open set whose measure is 2^{-n}. Identifying $^\omega 2$ with the binary expansions of reals in $[0, 1]$, this is Lebesgue measure. The topology is that of the Cantor set via the identification of $^\omega 2$ with the reals in $[0, 1]$ having a ternary representation containing only zeros and twos. We will also have occasion to consider the compact space $^A K$ where A is an arbitrary set and K is a finite set (with the discrete topology).

An *ideal* on a set A is a nonempty collection of subsets of A that is closed under finite unions and subset formation. An ideal is *non-principal* if it contains all singletons and *proper* if it does not contain the set A itself. Unless otherwise specified, when we say "ideal" we mean "proper, non-principal ideal."

Sets in an ideal I are said to be of I-*measure zero*; sets not in I of *positive I-measure*; and sets whose complement is in I of I-*measure one*. We let I^+ denote the collection of sets of positive I-measure and I^* denote the collection of sets of I-measure one. An important example of an ideal is $I = [\kappa]^{<\lambda}$ where $\lambda \leq \kappa$ are infinite cardinals.

A *filter* on a set A is a nonempty collection of subsets of A that is closed under finite intersections and superset formation. A filter is *non-principal* if it contains all cofinite sets and *proper* if it does not contain the empty set. Unless otherwise specified, when we say "filter" we mean "proper, non-principal filter." Notice that if I is an ideal, then I^* is a filter. A maximal filter is called an *ultrafilter*; it contains exactly one of X and $A - X$ for every $X \subseteq A$.

Finally, if I^* and J^* are filters on A then we can form the *product filter* $I^* \times J^*$ on $A \times A$ by $X \in I^* \times J^*$ iff $\{ a \in A : \{ b \in A : (a, b) \in X \} \in J^* \} \in I^*$.

1.3 Two Basic Negative Results

Throughout this section, we assume that A is the set of agents, K is the set of colors with $|K| \geq 2$, and V is a directed graph on A that serves as the visibility graph.

Definition 1.3.1. For any set X of agents, we define a sequence of subsets of X inductively as follows: $B_0 = \emptyset$, and for $0 < \alpha < |X|^+$, we let B_α consist of those agents $a \in X$ for which $V(a) \cap X \subseteq \cup\{ B_\beta : \beta < \alpha \}$. Let $B(X) = \cup\{ B_\alpha : \alpha < |X|^+ \}$.

Proposition 1.3.2. *For every predictor, every set X of agents, and every $f \in {}^{A-X} K$, there exists a hat coloring $f_X \in {}^A K$ that extends f and makes all agents in $B(X)$ guess incorrectly.*

Proof. Because of f, every agent in B_1 has his guess determined (in the sense that everyone they can see already has a hat placed on their head), so we can change the hats of those agents in B_1 who guessed correctly. Proceeding inductively, if we've placed hats on all the agents in $\cup\{B_\beta : \beta < \alpha\}$, then the guesses of all the agents in B_α have been determined so we can change hats where necessary to make them all guess incorrectly.   □

Corollary 1.3.3. *For every predictor, every set X of agents that is independent in V, and every $f \in {}^{A-X}K$, there exists a hat coloring $f_X \in {}^A K$ that extends f and makes all agents in X guess incorrectly.*

Proof. If X is independent in V, then $B_1(X) = \{a \in X : V(a) \cap X = \emptyset\} = X$. Thus $X = B_1(X) \subseteq B(X) \subseteq X$, so $B(X) = X$.   □

We later show (Theorem 2.2.2) that if there are two colors and the visibility graph has a cycle, then there is a predictor ensuring at least one agent guesses correctly. The necessity of the cycle in this context is shown by the following.

Corollary 1.3.4. *If the visibility graph is acyclic (i.e., has no cycles), then for every predictor, every finite set X of agents, and every $f \in {}^{A-X}K$, there exists a hat coloring $f_X \in {}^A K$ that extends f and makes all agents in X guess incorrectly.*

Proof. Suppose $|X| = n$. We can assign a rank to each $a \in X$ by saying that a has rank m if there is a directed path within X of length m beginning at a, but not one of length $m + 1$ beginning at a. Notice that if there is a directed edge from a to b, then the rank of a is strictly greater than the rank of b. Hence, each agent in X sees only agents in X of smaller rank. It now follows that $X \subseteq \cup\{B_k : k \le n\} \subseteq B(X) \subseteq X$.
□

To motivate our next definition and corollary, suppose that agent x cannot see y. Then for any predictor, we can make both x and y wrong by starting with an arbitrary coloring, changing x's hat to make x wrong, and then changing y's hat to make y wrong; since x cannot see the change to y, x's guess does not change so x and y are both now wrong. The following corollary is simply the extension of this idea to the transfinite, to get a simple upper bound on how well a predictor can possibly perform. We will see later that there are natural contexts in which predictors exist that attain this bound.

Definition 1.3.5. A set X of agents is *co-well-founded in V* if for every nonempty $X' \subseteq X$, there exists $x \in X'$ such that $V(x) \cap X' = \emptyset$.

Note that when V is a strict partial order, this agrees with the usual definition of co-well-founded.

Corollary 1.3.6. *For every predictor, every set X of agents that is co-well-founded in V, and every $f \in {}^{A-X}K$, there exists a hat coloring $f_X \in {}^A K$ that extends f and makes all agents in X guess incorrectly.*

Proof. We'll show that if X is co-well-founded in V, then $B(X) = X$. Suppose not and let $X' = X - B(X)$. Because $X' \neq \emptyset$ and X is co-well-founded, we can choose $x \in X'$ such that $V(x) \cap X' = \emptyset$. Thus $x \notin B(X)$ but $V(x) \cap X \subseteq B(X)$. Hence, for each $y \in V(x) \cap X$ we can choose $\alpha_y < |X|^+$ such that $y \in B_{\alpha_y}$. Let $\alpha = \sup\{\alpha_y : y \in V(x) \cap X\}$. Then $\alpha < |X|^+$ and $V(x) \cap X \subseteq \cup\{B_\beta : \beta < \alpha\}$. Hence, $x \in B_\alpha$, so $x \in B(X)$, a contradiction.   □

Our second technique here for producing negative results relies on the compactness of $^A2$.

Proposition 1.3.7. *If $|K| = 2$ and V is acyclic, then for every predictor, every set X of agents with the property that $V(x) \cap X$ is finite for every $x \in X$, and every $f \in {}^{A-X}K$, there exists a hat coloring $f_X \in {}^AK$ that extends f and makes all agents in X guess incorrectly.*

Proof. For each a, let I_a denote the set of 2-colorings that extend f and for which agent a guesses incorrectly using the given predictor. We make two claims.

Claim 1. For each $a \in A$, the set I_a is a closed subset of $^A2$.

Proof. Assume that $f_X \notin I_a$. If f_X disagrees with f at a, we need only consider the basic neighborhood corresponding to the set of all colorings that agree with f_X on $\{a\}$. Otherwise, we can consider the basic neighborhood given by the restriction of f_X to the set $V(a) \cup \{a\}$. If g is in this neighborhood, then agent a's guess with g is the same as his guess with f_X, which is correct. Hence $g \notin I_a$. This shows that I_a is closed.

Claim 2. The collection $\{I_a : a \in A\}$ has the finite intersection property.

Proof. Suppose $X = \{a_1, \ldots, a_k\}$, and let f_X be the hat coloring whose existence is guaranteed by Corollary 1.3.4. Then $f_X \in I_{a_1} \cap \cdots \cap I_{a_k}$ as desired.

With Claims 1 and 2, we can now apply the compactness of $^A2$ to get a coloring f in all of the I_as.   □

It is easy to see that Proposition 1.3.7 holds as long as K is a finite set with two or more elements.

1.4 One Basic Positive Result: The μ-Predictor

This section contains only one definition (and no lemmas, propositions, or theorems). But the definition given here is—at least for the purposes of the present monograph—the most important one that we present. The μ-predictor was first introduced in [HT08b].

Definition 1.4.1. Suppose A is the set of agents, K is the set of colors, and $\equiv_a$ is agent a's equivalence relation on $^A K$. Let $[f]_a$ denote the equivalence class of f with respect to $\equiv_a$. We now fix a well-ordering $\prec$ of $^A K$ and, for each $a \in A$, we let $\langle f \rangle_a$ denote the $\prec$-least element of $[f]_a$. Agent a's guessing strategy, denoted μ_a, is to assume the given coloring is the $\prec$-least one that is consistent with what he can see. More precisely, $\mu_a(f) = \langle f \rangle_a(a)$. The predictor $M_\prec = \langle \mu_a : a \in A \rangle$ is called the μ-*predictor*, and thus $M_\prec(f)(a) = \langle f \rangle_a(a)$ for every $a \in A$.

One can remember the way the μ-predictor works more easily by noting the similarity to Occam's razor, which would have us guess according to the simplest theory consistent with what can be observed. The μ-predictor first occurs in Sect. 4.2; the reader wishing to see an example of its use can glance ahead at the proof of Theorem 4.2.1. Another example is Theorem 7.3.3, which complements Corollary 1.3.6 above by showing that if V is a strict partial order, then the set of agents who guess incorrectly using the μ-predictor is co-well-founded.

In a great many situations in later chapters, when we need a successful predictor, we turn to some variant of the μ-predictor.

1.5   A Preview of What Is to Come

We begin in Chap. 2 with the case in which the set A of agents is finite and the equivalence relations $\equiv_a$ are given by a directed graph V on A. So how much visibility is needed for the existence of a minimal predictor? And for a given visibility graph, what is the value of k for an optimal predictor? These are the kinds of questions dealt with in Chap. 2.

The context in Chap. 3 is similar to that in Chap. 2 except that we focus on the case in which the set of agents is denumerable. Results from the finite setting suggest there should be a predictor ensuring that infinitely many agent guess correctly, and indeed, this turns out to be trivial. But something quite unexpected also turns out to be true, as was noticed by Yuval Gabay and Michael O'Connor while they were graduate students at Cornell in 2004. Regardless of how large the infinite set of agents is and regardless of how large the (perhaps infinite) set of hat colors is, there exists a predictor ensuring that only finitely many agents guess incorrectly. We are assuming here that every agent sees all the hats but his own. And we are most definitely assuming the axiom of choice.

In Chap. 4 we restrict ourselves to "one-way visibility" on ω. This means that agents never see smaller-numbered agents. In this context, visibility can be given by an undirected graph, wherein an edge means the smaller vertex can see the larger. We introduce here a highly nontransitive setting that corresponds to the agents being natural numbers with each even agent seeing all the higher-numbered odd agents and vice-versa. Quite unexpectedly, it turns out that the existence of a minimal predictor here is very dependent on the number of colors. In fact, with finitely many colors, there always is one, while with $\aleph_2$ colors, there never is one. The continuum hypothesis yields one if the are $\aleph_1$ colors, and the existence of one with a denumerable set of colors is independent of ZFC $+2^{\aleph_0} = \aleph_2$. It also turns out that P-point and Ramsey ultrafilters arise naturally in this context.

In Chap. 5 we study dual hat problems where, roughly speaking, notions corresponding to injective functions or subsets are altered by considering surjective functions or partitions. Within the hat problem metaphor this shifts the focus from near-sightedness to colorblindness. We then move on to ideals on both countable and uncountable sets and we see the role played by non-regular ultrafilters. We also establish an equivalence with the GCH.

The context originally studied by Galvin in the 1960s involved predictors for ω where (roughly speaking) the agents did not know where in line they were standing. Such considerations lead naturally to so-called "neutral" and "anonymous" predictors, and these are investigated in Chap. 6.

In Chap. 7 we start to move further away from the hat problem metaphor and think instead of trying to predict a function's value at a point based on knowing (something about) its values on nearby points. The most natural setting for this is a topological space and if we wanted to only consider *continuous* colorings, then the limit operator would serve as a unique optimal predictor. But we want to consider arbitrary colorings. Thus we have each point in a topological space representing an agent and if f and g are two colorings, then $f \equiv_a g$ if f and g agree on some deleted neighborhood of the point a. It turns out that an optimal predictor in this case is wrong only on a set that is "scattered" (a concept with origins going back to Cantor).

To illustrate one corollary of this topological result, consider the hat problem in which the agents are indexed by real numbers, and each agent sees the hats worn by those to his left (that is, those indexed by smaller real numbers). The set of hat colors is some arbitrary set K. The question is whether or not there is a predictor ensuring that the set of agents guessing incorrectly is a small infinite set—e.g., indexed by a set of reals that is countable and nowhere dense. The answer here (again assuming the axiom of choice) is yes. In fact, there is a predictor guaranteeing the set of agents guessing incorrectly is a set of reals that is well ordered by the usual ordering of the reals. Moreover, this is an optimal predictor.

We published the above result in the *American Mathematical Monthly* in an article with the deliberately provocative title "A peculiar connection between the axiom of choice and predicting the future." The point is that this hat problem allows one to conclude that if time is modeled by the real line, then the axiom of choice guarantees that the present state of any system can "almost always" be correctly predicted based on its past states. The economist Steven Landsburg, author of *The Armchair Economist*, (The Free Press, 1993), referred to this in his internet blog as "the single most counterintuitive-but-true thing I've ever heard." While this is almost certainly an overstatement, the result is reminiscent of the Banach-Tarski paradox, which allows one to conclude that if space is modeled by $\mathbf{R}^3$, then the axiom of choice guarantees that a solid ball can be divided into finitely many parts and reassembled by rotations and translations into two balls identical to the original solid ball.

In Chap. 8 we show that the scattered-error predictor from Chap. 7 is essentially unique: in many contexts, all predictors that perform as well as the μ-predictor are special cases of the μ-predictor. Finally, in Chap. 9, we explore the relationships between results extending those in Chap. 4 and so-called Galois-Tukey connections.

The authors are indebted to three anonymous referees for a number of suggestions that substantially improved the presentation, and to our editor at Springer-Verlag, Elizabeth Loew.

Chapter 2
The Finite Setting

2.1 Background

Although our primary interest in this monograph is with the infinite, we begin with a discussion of hat problems in which the set A of agents is finite and visibility is given by a directed graph V on A (the visibility graph). The set K of colors is typically a natural number k and the set C of colorings is the entire set $^A K$. Thus if f and g are two colorings in $^A K$ and a is an agent, then $f \equiv_a g$ iff $f(b) = g(b)$ for every $b \in V(a)$.

As indicated in the introduction, we think of each agent a as trying to guess his own hat color via a guessing strategy $G_a : C \to K$ satisfying $G_a(f) = G_a(g)$ whenever $f \equiv_a g$. The predictor P arising from these guessing strategies is given by $P(f)(a) = G_a(f)$, and agent a guesses correctly for f if $G_a(f) = f(a)$; equivalently, if $P(f)(a) = f(a)$.

Most of what is known in the finite case (where agents cannot pass) can be found in a single paper entitled *Hat Guessing Games*, by Steven Butler, Mohammed Hajiaghayi, Robert Kleinberg, and F. Thomson Leighton [BHKL08]. Some of these results were obtained independently and appeared in [HT08a].

The rest of this chapter is organized as follows. In Sect. 2.2 we illustrate the notion of a minimal predictor, including a result from [BHKL08] on bipartite visibility graphs that we will need in Chap. 4. In Sect. 2.3, we consider optimal strategies, and in Sect. 2.4 we present another result from [BHKL08] that uses the Tutte-Berge formula to completely solve the prediction problem for finite symmetric graphs. In Sect. 2.5 we consider the situation in which the agents have different color-sets for their hats, and in Sect. 2.6 we look at some variants of the standard hat problem, one of which will arise again in Chap. 3. We conclude in Sect. 2.7 with a discussion of some open questions.

C.S. Hardin and A.D. Taylor, *The Mathematics of Coordinated Inference*,
Developments in Mathematics 33, DOI 10.1007/978-3-319-01333-6_2,
© Springer International Publishing Switzerland 2013

2.2   Minimal Predictors

In the present context, a minimal predictor will be one that guarantees at least one correct guess. Thus, with n agents, we can ask how much visibility is needed for the existence of a minimal predictor. Stated differently, for a fixed number of colors, we seek a characterization of those visibility graphs that yield a minimal predictor. Our first theorem answers this for the case of two colors and the case of n colors; the result appears as Theorem 1 in [HT08a], although most of it can be derived from results in [BHKL08]. But first we need a lemma that confirms an intuition about how many agents guess correctly on average.

Lemma 2.2.1. *In an n-agent, k-color hat problem, for any particular predictor, the average number of agents who guess correctly is n/k. (The average is taken over all colorings.)*

Proof. Suppose there are n agents and k colors. Let P be any predictor. It suffices to show that any particular agent a is correct in one out of k colorings. Given any assignment of hat colors to all agents other than a, agent a's guess will be determined; of the k ways to extend this hat coloring to a, exactly one will agree with a's guess.   □

Theorem 2.2.2. *For an n-agent, 2-color hat problem, there is a predictor ensuring at least one correct guess iff the visibility graph has a cycle. For an n-agent, n-color hat problem, there is a predictor ensuring at least one correct guess iff the visibility graph is complete.*

Proof. Suppose first that there are two colors. For the right-to-left direction, assume the visibility graph has a cycle. Fix an agent on the cycle and let his strategy be to guess assuming his hat and the one on the agent just ahead of him on the cycle are the same color. The other agents on the cycle guess according to the opposite assumption; they assume that their hat and that of the agent just ahead of them on the cycle are different colors. To see that this works, assume that the first agent on the cycle has a red hat and that everyone on the cycle guesses incorrectly using this strategy. Then the second agent on the cycle has a green hat, the third agent on the cycle has a green hat, and so on until we're forced to conclude that the first agent on the cycle also has a green hat, contrary to what we assumed.

For the other direction, we can appeal to Corollary 1.3.4 which ensures that if there is no cycle in the visibility graph V, then for every predictor there is a coloring for which everyone guesses incorrectly.

Now suppose there are n colors. For the right-to-left direction, assuming the visibility graph is complete, the strategies are as follows. Number the agents $0, 1, \ldots, n-1$, and the colors $0, 1, \ldots, n-1$, and for each i, let s_i be the mod n sum of the hats seen by agent i. The plan is to have agent i guess $i - s_i \pmod{n}$ as the color of his hat. If the colors of all the hats add to $i \pmod{n}$, then agent i will be the one who guesses correctly. That is, if $c_0 + \cdots + c_{n-1} = i \pmod{n}$ then $c_i = i - s_i \pmod{n}$.

For the other direction, assume that there are n agents and n colors, and assume the visibility graph is not complete. Let P be any predictor. We must show that there is a coloring in which every agent guesses incorrectly. Suppose agent a does not see agent b's hat (with $a \neq b$), and pick a coloring in which agent a guesses correctly. If we change the color of agent b's hat to match agent b's guess, agent a will not change his guess, and we will have a coloring in which a and b guess correctly. By Lemma 2.2.1, the average number of agents who guess correctly is $n/n = 1$; because we have a coloring with at least two agents guessing correctly, there must be another coloring in which less than one (namely, zero) agents guess correctly. □

We conclude this section with one other result about minimal predictors. It was first established in [BHKL08] and later rediscovered by Daniel J. Velleman [Vel11] in his solution to a question left open in [HT10]. Velleman's result is in Chap. 4.

Theorem 2.2.3. *For every k there is a finite bipartite graph V such that if visibility is given by V with agent a seeing agent b only if a is adjacent to b, then there is a predictor for the k-color hat problem ensuring at least one correct guess.*

Proof. Thinking of the problem as involving two teams that see each other, the trick is to get team 1 to guess in such a way that if they're all wrong, there are relatively few possibilities for how team 2 is colored.

Let team 2 have $k - 1$ agents. There are only finitely many possible individual strategies for agents on team 1, so let team 1 have one agent for each possible individual strategy. We claim that given a coloring f of team 1 and k distinct colorings $g_1, \ldots, g_k$ of team 2, someone on team 1 is correct for at least one of these colorings; for, at least one of the agents on team 1 will guess a different color for each of $g_1, \ldots, g_k$, so one of these guesses must agree with f. In light of that claim, team 2 does the following: assuming every guess on team 1 is wrong leaves at most $k - 1$ different possible colorings of team 2, so with $k - 1$ agents, they can let agent i among them guess according to the ith such possibility. □

Note that with k colors, both teams need at least $k - 1$ agents, as the following argument shows. Suppose team 2 has fewer than $k - 1$ agents, and that team 1 has any set (possibly infinite) of agents. Fix any predictor P. We will commit to coloring team 2 with a constant coloring among $0, \ldots, k - 2$. Let f_i be the coloring guessed by team 1 when team 2 is all colored i, for $i = 0, \ldots, k - 2$. There are only $k - 1$ such f_is, so for each agent a on team 1, there is a color that differs from $f_0(a), \ldots, f_{k-2}(a)$, so we can fix a coloring h of team 1 that makes them all wrong when team 2 is all colored i, for $0 \leq i \leq k - 2$. Now, look at what team 2 guesses when they see h. Since they have only $k - 2$ agents, there is some color among $0, \ldots, k - 2$ that none of them guesses; color everyone on team 2 this color.

2.3   Optimal Predictors

As we said in the preface, an optimal predictor achieves a degree of correctness that is maximal in some sense. For the case of a visibility graph that is complete, there is a very satisfying result that we present below. It occurs as Theorem 2 in [BHKL08] and as Theorem 3 in [HT08a], although it was first proved for two colors by Peter Winkler [Win01] and later generalized to k colors by Uriel Feige [Fei04].

By way of motivation, recall that Lemma 2.2.1 showed that, regardless of strategy, if there are n agents and k colors, the number who guess correctly will *on average* be n/k. But this is very different from ensuring that a certain fraction will guess correctly regardless of luck or the particular coloring at hand. Nevertheless, the fraction n/k is essentially the correct answer.

Theorem 2.3.1. *Consider the hat problem with* $|A| = n$, $|K| = k$, *and a complete visibility graph* V. *Then there exists a predictor ensuring that* $\lfloor n/k \rfloor$ *agents guess correctly, but there is no predictor ensuring that* $\lfloor n/k \rfloor + 1$ *agents guess correctly.*

Proof. The strategy ensuring that $\lfloor n/k \rfloor$ agents guess correctly is obtained as follows. Choose $k \times \lfloor n/k \rfloor$ of the agents (ignoring the rest) and divide them into $\lfloor n/k \rfloor$ pairwise disjoint groups of size k. Regarding each of the groups as a k-agent, k-color hat problem, we can apply Theorem 2.2.2 to get a strategy for each group ensuring that (precisely) one in each group guesses correctly. This yields $\lfloor n/k \rfloor$ correct guesses altogether, as desired.

For the second part, we use Lemma 2.2.1. For any predictor, the average number of agents who guess correctly will be n/k, and $n/k < \lfloor n/k \rfloor + 1$, so no predictor can guarantee at least $\lfloor n/k \rfloor + 1$ agents guess correctly for each coloring.   □

2.4   The Role of the Tutte-Berge Formula

In this section, we consider only symmetric visibility graphs on a finite set A of agents and we present a very nice result from [BHKL08] that specifies exactly how successful a predictor for two colors can be for such a visibility graph. The obvious strategy with a symmetric visibility graph $V = (A, E)$ is to pair up agents who can see each other, and then have them use the trivial two-agent strategy that we described in Chap. 1. Remarkably, this obvious strategy turns out to be optimal.

A *matching* M for a graph V is a collection of pairwise disjoint edges, where we are thinking of an edge as a two-element set. Thus, the *size* of the matching M is literally the cardinality of M; that is, the number of edges in the matching. A vertex is said to be *covered* by the matching M if it is in the union of M, that is, if it is an endpoint of one of the edges in M. Thus, if M is a matching of maximum size for a finite symmetric visibility graph, then there is a two-color predictor ensuring that at least $|M|$ agents guess correctly. The theorem below shows that no predictor can ensure more.

First, however, we need to discuss the so-called Tutte-Berge formula for the maximum size of a matching for a graph V. William Tutte's original contribution [Tu47] was in characterizing those graphs V for which there is a matching that covers every vertex of V. His starting point was with an obvious necessary condition for such a matching: Every set $S \subseteq A$ of vertices must have at least as many points as there are odd-sized components in $A - S$. The point is that in a component of $V - S$, vertices can appear in the matching only when paired with either another element of that component, or with a vertex in S. If the component has odd size, at least one of the vertices in the component will need to be paired with an element of S. And different components require different vertices in S. Tutte showed that this necessary condition for a matching covering all vertices of V was also sufficient.

Claude Berge's generalization [B58] of Tutte's result involves the finer analysis resulting from the "deficiency" of a set S in having enough vertices to handle each of the leftover vertices in the components of odd size in $V - S$. Notationally, let $\mathcal{O}(V - S)$ denote the set of components of odd size in $V - S$, and let $\text{def}(S)$, the *deficiency of S*, be given by $\text{def}(S) = |\mathcal{O}(V - S)| - |S|$.

It now follows from what we've said that for every set $S \subseteq V$, every matching will leave at least $\text{def}(S)$ vertices not covered. Thus, if M is a matching of maximum size, then for every set $S \subseteq V$, we can cover at most $V - \text{def}(S)$ vertices, and so we must have $|M| \leq \frac{1}{2}(|V| - \text{def}(S))$. Berge's contribution was to show that equality always holds for some $S \subseteq V$. This is the Tutte-Berge formula.

With this at hand, we can now establish the following from [BHKL08].

Theorem 2.4.1. *Let V be any finite graph and consider the corresponding two-color hat problem with symmetric visibility given by V. Let M be a matching of maximum size for V. Then there is a predictor ensuring $|M|$ correct guesses, and there is no predictor ensuring $|M| + 1$ correct guesses.*

Proof. The predictor ensuring $|M|$ correct guesses is the one described in the first paragraph of this section. What must be shown is that no predictor P can do better. So let S be a set of agents (vertices) as in the Tutte-Berge formula, wherein $|M| \leq \frac{1}{2}(|V| - \text{def}(S))$. Let $W_1, \ldots, W_j$ denote the components of odd size in $V - S$, and let $Y = (V - S) - (W_1 \cup \cdots \cup W_j)$. Thus A is the disjoint union of S, Y, and the W_is. We begin by placing hats on S arbitrarily. Any agent in W_i sees only other agents in W_i or agents in S. Because hats have been placed on agents in S, we can regard the predictor P as operating on W_1 alone, and by Theorem 2.3.1 we can place hats so as to make at most $\frac{1}{2}(|W_1| - 1)$ of the agents in W_1 guess correctly. We do this for each W_i. Finally, we place hats on the agents in Y so that at most half of them guess correctly. It now follows that the total number of agents guessing correctly for this hat coloring is at most

$$|S| + \frac{1}{2}(|W_1| - 1) + \cdots$$

$$+ \frac{1}{2}(|W_j| - 1) + \frac{1}{2}|Y| = |S| + \frac{1}{2}(|W_1| + \cdots + |W_j| - j + |Y|)$$

$$= |S| + \frac{1}{2}(|V - S| - j)$$

$$= \frac{1}{2}|S| + \frac{1}{2}|V - S| + \frac{1}{2}|S| - \frac{1}{2}j$$

$$= \frac{1}{2}|V| + \frac{1}{2}|S| - \frac{1}{2}j$$

$$= \frac{1}{2}(|V| - \text{def}(S))$$

$$= |M|. \qquad\qquad \square$$

2.5   A Variable Number of Hat Colors

Suppose we have finitely many agents, each of whom can see all of the others. When does there exist a minimal predictor? If the same set of colors is used for each agent, we already know: there is a minimal predictor iff there are at least as many agents as colors. But what if different agents have different sizes of sets from which their hat colors are drawn?

Suppose then that $p = \{0, \ldots, p-1\}$ is our set of agents and that $c_0, \ldots, c_{p-1} \in \omega - \{0\}$. We will assume that agent i's hat color will be in the set $c_i = \{0, \ldots, c_i - 1\}$, and we will let the tuple $(c_0, c_1, \ldots, c_{p-1})$ encode this problem. Our first observation is that if $\sum_i 1/c_i < 1$, then there is no minimal predictor because, as in Lemma 2.2.1, the average number of correct guesses will be less than one. A natural question here is whether or not the converse holds. That is, if $\sum_i 1/c_i \geq 1$, must there be a minimal predictor? The following theorem shows that the answer is yes.

Theorem 2.5.1. *Let $p \in \omega$ and $c_0, \ldots, c_{p-1} \in \omega - \{0\}$, and consider the hat problem in which p is the set of agents and the set of colorings is $\{ f \in {}^p\omega : (\forall i \in p)(f(i) \in c_i) \}$; that is, agent i has c_i possible hat colors. Let $r = \sum_i 1/c_i$. Provided no agent sees himself, the average number of correct guesses will be r, regardless of the predictor. If every agent sees every other agent, then there is a predictor under which the number of correct guesses is always $\lfloor r \rfloor$ or $\lceil r \rceil$. In particular, a predictor ensuring at least one correct guess exists iff $r \geq 1$.*

Proof. The fact that the average number of correct guesses will be r is again by the same argument as in Lemma 2.2.1.

Suppose now that every agent sees every other agent. We define the predictor P as follows. Colorings can be seen as elements of the group $C = \mathbf{Z}_{c_0} \oplus \cdots \oplus \mathbf{Z}_{c_{p-1}}$ in the obvious fashion. Define $\pi : \mathbf{R} \to \mathbf{R}/\mathbf{Z}$ (as a group homomorphism) by $\pi(x) = x + \mathbf{Z}$ (that is, we are projecting modulo $\mathbf{Z}$). Define $\varphi : C \to \mathbf{R}/\mathbf{Z}$ by

$$\varphi(f) = \pi\left(\sum_{k \in p} \frac{f(k)}{c_k} \right).$$

Define $d_k \in \mathbf{R}$ by $\sum_{j<k} 1/c_j$, let $\hat{I}_k = [d_k, d_k + 1/c_k) \subseteq \mathbf{R}$, and let $I_k = \pi[\hat{I}_k]$. The intervals $\hat{I}_k$ lie end-to-end, and have total length r, so when we project to $\mathbf{R}/\mathbf{Z}$, each point in $\mathbf{R}/\mathbf{Z}$ occurs in $\lfloor r \rfloor$ or $\lceil r \rceil$ of the I_k.

For our predictor, agent k assumes that the coloring f satisfies $\varphi(f) \in I_k$ and guesses accordingly. (This is a well-defined predictor: agent k knows the value of $\varphi(f)$ up to a multiple of $1/c_k$, and exactly one of these multiples would put $\varphi(f)$ in I_k, because I_k is left-closed right-open with length $1/c_k$.) Now, agent k will guess correctly iff $\varphi(f) \in I_k$, and as observed above, this occurs for $\lfloor r \rfloor$ or $\lceil r \rceil$ values of k.

When $r \geq 1$, of course, the number of correct guesses under this predictor is at least one, so we have a minimal predictor; when $r < 1$, the average number of correct guesses is less than one, so there can be no minimal predictor. $\square$

2.6   Variations on the Standard Hat Problem

In addition to the kind of hat problem we are considering, several interesting variants have arisen over the years. We'll mention two of these here.

The first variant is the following. Ten prisoners are lined up facing forward, and red and green hats are placed on their heads. Each prisoner will be asked to make a verbal guess as to the color of his hat, and, before guessing, each will be able to hear the guesses of those prisoners behind him as well as seeing the hats of those prisoners in front of him. If at most one guesses incorrectly, all will go free.

No strategy can ensure that the first prisoner will guess correctly, but if the first player uses his guess to announce red iff he sees an even number of red hats, then all of the others can use this signal (and the knowledge that the others are also using it) to correctly guess their hat color.

The extension of this problem and its solution to countably many agents will be given in Chap. 3. We also show there that these signaling problems are equivalent (in ZF + DC) to the kind of non-signaling problems that we are considering.

The second variant goes as follows. There are again ten prisoners, this time wearing shirts that are numbered one through ten. Each prisoner has a hat with a number on it matching the number on his shirt. The warden confiscates the hats and places them randomly in boxes numbered one through ten. One-by-one, the prisoners are called to the room and allowed to open nine of the ten boxes. If all ten prisoners find their own hats, then all go free. If any one of the ten fails, they all remain in prison. Find a strategy yielding a 90% chance that all will go free.

The solution is for prisoner i to begin by looking in box i. If he sees hat j, then he next looks in box j. And so on. The only way for any prisoner to lose using this strategy is for the placement of the hats to correspond to one of the 9! ways to place 10 numbers around a circle. But there are 10! permutations, so the chance of failure is only 9!/10! = 1/10.

2.7   Open Questions

There is no shortage of questions that could be stated here, but we'll introduce a bit of notation in order to state the most obvious (and perhaps the most difficult). For positive integers k, n, and m, let $P_k(n, m)$ denote the collection of all directed graphs on n vertices ("n-graphs") for which there is a predictor ensuring that at least m agents guess correctly when there are k hat colors. The results in this section show that:

- $P_2(n, 1)$ is the collection of n-graphs with a cycle.
- $P_n(n, 1)$ is the collection of n-graphs that are complete.
- $P_k(n, \lfloor n/k \rfloor + 1)$ is the empty collection.

Question 2.7.1. Can one characterize the graphs in $P_k(n, m)$ for other values of k, n, and m?

There are also two questions related to the theorem on bipartite graphs; the first is from [BHKL08], and the second is due to Velleman.

Question 2.7.2. Is there a bipartite graph ensuring the existence of a minimal predictor for k colors whose size is polynomial in k?

Question 2.7.3. Given cardinals (possibly finite, treated as sets) c, m, k, with $k \le c$, what is the smallest size of a family F of functions from m to c such that, for every subset A of m of size k, $f \mid A$ is one-to-one for some $f \in F$?

There is an old saying, variously attributed to everyone from the French Minister Charles Alexandre de Calonne (1734–1802) to the singer Billie Holiday (1915–1959), that goes roughly as follows: "The difficult is done at once; the impossible takes a little longer." More to the point, Stanislaw Ulam (1909–1984) provided the adaptation that says, "The infinite we shall do right away; the finite may take a little longer." With this in mind, we leave the finite.

Chapter 3
The Denumerable Setting: Full Visibility

3.1 Background

With two colors and an even number of agents, Theorem 2.3.1 says that—with collective strategizing—the on-average result of 50 % guessing correctly can, in fact, be achieved with each and every coloring. But it also says that this is the best that can be done by collective strategizing. In the finite case, this latter observation does little more than provide proof for what our intuition suggests: collective strategizing notwithstanding, the on-average result of 50 % cannot be improved in a context wherein guesses are simultaneous. The infinite, however, is very different, and it is to this that we next turn.

With two colors, it is easy to produce a predictor for a denumerably infinite set of agents ensuring infinitely many will guess correctly. One can, for example, pair up the agents and let each pair use the strategy given in Sect. 1.1. Or, if agents only see higher-numbered agents, one can have an agent guess red if he sees infinitely many red hats, and guess green otherwise. If there are infinitely many red hats, everyone will guess red and the agents with red hats will be correct; if there are finitely many red hats, everyone will guess green, and the cofinitely many agents with green hats will be correct. This generalizes to the case of a finite set of colors by numbering the colors and having each agent guess that his hat color is the lowest numbered color that occurs infinitely often.

But as we have said, something much more striking is true. There is a predictor ensuring that all but finitely many—not just infinitely many—are correct, and this is what Gabay and O'Connor obtained using the axiom of choice. Such a predictor is called a *finite-error predictor*. In fact, with the Gabay-O'Connor theorem, the set of agents and the set of colors can be arbitrary, although the special case in which the set of agents is countable follows from a 1964 result of Galvin (see [Gal65] and [Tho67]). While Galvin's argument and the Gabay-O'Connor argument are similar, they are different enough that neither subsumes the other.

Throughout this chapter, we take the set A of agents to be the set ω, and if there are two colors, we take them to be 0 and 1.

C.S. Hardin and A.D. Taylor, *The Mathematics of Coordinated Inference*,
Developments in Mathematics 33, DOI 10.1007/978-3-319-01333-6_3,
© Springer International Publishing Switzerland 2013

The rest of this chapter is organized as follows. In Sect. 3.2, we present the Gabay-O'Connor theorem and we point out the essential uniqueness of their predictor. In Sect. 3.3, we derive from this a result of Hendrik Lenstra that guarantees a predictor ensuring that every agent will guess correctly or every agent will guess incorrectly. We also consider the kind of sequential guessing that arose in Chap. 2, and we prove an equivalence between this and Lenstra's context. Sections 3.4 and 3.5 examine the need for the axiom of choice, with the former relying on the property of Baire and the latter relying on square bracket partition relations. Some open questions are given in Sect. 3.6.

3.2   The Gabay-O'Connor Theorem

We begin with a statement and proof of the Gabay-O'Connor theorem, but this requires one definition.

Definition 3.2.1. A predictor P is *robust* if each agent's guess is unchanged if the colors of finitely many hats are changed; that is, P is robust if $P(f) = P(g)$ whenever $f \Delta g$ is finite. More generally, for an ideal I on the set of agents, P is *I-robust* if $P(f) = P(g)$ whenever $f \Delta g \in I$.

Theorem 3.2.2 (Gabay, O'Connor). *Consider the situation in which the set A of agents is arbitrary, the set K of colors is arbitrary, and every agent sees all but finitely many of the other hats. Then there exists a predictor ensuring that all but finitely many agents guess correctly. Moreover, the predictor is robust.*

Proof. For $h, g \in {}^A K$, say $h =^* g$ if $h \Delta g$ is finite; this is an equivalence relation on ${}^A K$. By the axiom of choice, there exists a function $\Phi : {}^A K \to {}^A K$ such that $\Phi(h) =^* h$ and if $h =^* g$, then $\Phi(h) = \Phi(g)$. Thus, Φ is choosing a representative from each equivalence class. Notice that for each coloring h, each agent a knows the equivalence class $[h]$, and thus $\Phi(h)$, because the agent can see all but finitely many hats. The strategies are then to have the agents guess their hat colors according to the chosen representative of the equivalence class of the coloring; more formally, we are letting $G_a(h) = \Phi(h)(a)$. For any coloring h, since this representative $\Phi(h)$ only differs from h in finitely many places, all but finitely many agents will guess correctly. Also, if finitely many hats change colors, the equivalence class remains the same and agents keep the same guesses.   □

Theorem 3.2.2 is sharp in the sense that even with countably many agents and two colors, no predictor can ensure that, for a fixed m, all but m agents will guess correctly, even if everyone sees everyone else's hat. The reason is that any such predictor would immediately yield a strategy for $2m + 1$ agents in which more than 50 % would guess correctly each time, contradicting Lemma 2.2.1.

The robust predictor whose existence is given by the Gabay-O'Connor theorem is essentially unique. That is, if $P = \langle G_a : a \in A \rangle$ is a finite-error predictor that is robust, then there exists a function $\Phi : {}^A K \to {}^A K$ such that $\Phi(h) \approx h$ and if $h \approx g$, then $\Phi(h) = \Phi(g)$—and such that $G_a(h) = \Phi(h)(a)$. In fact, the function Φ is just P itself. To see this, note first that $P(h) \approx h$ because P is a finite-error predictor. Moreover, because P is robust, we have that if $f \approx g$ then $P(f) = P(g)$. And finally $G_a(h) = P(h)(a)$, by definition of P in terms of the G_as. It is also worth noting that for any f, $P(f)$ is the unique fixed point in f's equivalence class; that is, we have $f \approx P(f)$, and so $P(f) = P(P(f))$.

The proof of the Gabay-O'Connor theorem above naturally extends to the following more general form. (Recall the definition of product filter from page 4.)

Theorem 3.2.3 (Gabay, O'Connor). *Let I be an ideal on A, and suppose we have a visibility graph $V \in I^* \times I^*$; that is, an I-measure one set of agents sees an I-measure one set of agents. Let the set K of colors be arbitrary. Then there exists an I-robust predictor guaranteeing that the set of agents guessing incorrectly is in I.*

Proof. Replacing all references to finiteness in the proof of Theorem 3.2.2 with membership in I, we are almost done. The only catch is that some agents a may have $V(a) \notin I^*$; we let these agents guess arbitrarily, and since the set of such agents is in I, even if they all guess incorrectly, the set of agents guessing incorrectly is still in I. □

3.3 Lenstra's Theorem and Sequential Guessing

The following theorem was originally obtained by Hendrik Lenstra using techniques (described below) quite different from our derivation of it here from Theorem 3.2.2.

Theorem 3.3.1 (Lenstra). *Consider the situation in which the set A of agents is arbitrary, $|K| = 2$, and every agent sees all of the other hats. Then there exists a predictor under which everyone's guess is right or everyone's guess is wrong.*

Proof. Let P be the predictor in Theorem 3.2.2. A useful consequence of the robustness of P is that, for a given coloring h, an agent a can determine $G_b(h)$ for every agent b. Since we are assuming agents can see *all* other hats, a also knows the value of $h(b)$ for every $b \neq a$. So, we can define a predictor T by letting $T_a(h) = G_a(h)$ iff $|\{ b \in P : b \neq a \text{ and } G_b(h) \neq h(b) \}|$ is an even number. That is, the agents take it on faith that, when playing P, an even number of agents are wrong: if they see an even number of errors by others, they keep the guess given by P, and otherwise they switch.

To see that T works, let h be a given coloring. When $|\{ b \in P : G_b(h) \neq h(b) \}|$ is even, every guess given by T will be correct: the agents who were already correct under P will see an even number of errors (under P), and keep their guess; the agents who were wrong under P will see an odd number of errors and switch. When $|\{ b \in A : G_b(h) \neq h(b) \}|$ is odd, the opposite occurs, and every guess given by T will be incorrect: the agents who would be correct under P will see an odd number of errors and switch (to the incorrect guess); the agents who would be wrong under P will see an even number of errors and stay (with the incorrect guess).   □

The assumption that everyone can see everyone else's hat in Theorem 3.3.1 is necessary. That is, if agent a could not see agent b's hat, then changing agent b's hat would change neither his nor agent a's guess, but agent b would go from wrong to right or vice-versa, and agent a would not.

Lenstra's theorem can be generalized from two colors to the case in which the set of colors is an arbitrary (even infinite) Abelian group. The conclusion is then that, for a given coloring, everyone's guess will differ from their true hat color by the same element of the group. Intuitively, the strategy is for everyone to take it on faith that the (finite) group sum of the differences between the true coloring and the guesses provided by the Gabay-O'Connor theorem is the identity of the group. (Variants of this observation were made independently by a number of people.)

Lenstra's original proof is certainly not without its charms, and goes as follows. If we identify the color red with the number zero and the color green with the number one, then we can regard the collection of all colorings as a vector space over the two-element field. The collection W of all colorings with only finitely many red hats is a subspace, and the function that takes each such coloring to zero if the number of red hats is even, and one otherwise, is a linear functional defined on W. The axiom of choice guarantees that this linear functional can be extended to the whole vector space. Moreover, a coloring is in the kernel iff the changing of one hat yields a coloring that is not in the kernel. Hence, the strategy is for each agent to guess his hat color assuming that the coloring is in the kernel. If the coloring is, indeed, in the kernel, then everyone guesses correctly. If not, then everyone guesses incorrectly.

Another proof of Lenstra's Theorem, at least for the case where the set of players is countably infinite, was found by Stan Wagon. It uses the existence, ensured by the axiom of choice, of a (non-principal) ultrafilter on A. Wagon's proof goes as follows. Label the agents by natural numbers and call an integer a "red-even" if the number of red hats among agents $0, 1, \ldots, a$ is even. Agent a's hat color affects which integers $b > a$ are red-even in the sense that changing agent a's hat color causes the set of red-even numbers greater than a to be complemented. The strategy is for agent a to make his choice so that, if this choice is correct, then the set of red-even numbers is in the ultrafilter $\mathcal{U}$. The strategy works because either the set of red-even numbers is in $\mathcal{U}$ (in which case everyone is right) or the set of red-even numbers is not in $\mathcal{U}$ (in which case everyone is wrong).

In Sect. 2.6, we described the 10-prisoner hat problem in which the prisoners are lined up facing forward so that each sees all the hats ahead of him, and the guesses are sequential so that each prisoner also hears the guesses of all the agents behind him before venturing his own guess. The extension of this problem and its solution to countably many players (described below) is due to Yuval Gabay. Our interest here is in the ways these signaling problems are equivalent (in ZF + DC) to the kind of non-signaling problem in Lenstra's theorem. The following is one example.

Theorem 3.3.2 (ZF + DC). *With an arbitrary (finite or infinite) set A of agents, an arbitrary visibility graph V, and two colors, the following are equivalent:*

1. *The visibility graph V is complete and there exists a "signaling strategy" S under which a designated agent ("agent 0") guesses his hat color out loud (this guess may or may not be correct—it should be thought of more as a signal than a guess), and then everyone else simultaneously produces a correct guess of his own hat color.*
2. *There exists a predictor P under which everyone simultaneously guesses his own hat color, and either everyone guesses correctly or everyone guesses incorrectly.*

Proof. Assume that V is complete, $K = 2$, and that $S = \langle S_a : a \in A \rangle$ is a signaling strategy as in (1). Note the following (the last of which we will prove momentarily):

(a) $S_0 : {}^A K \to K$ and if $h | A - \{0\} = g | A - \{0\}$, then $S_0(h) = S_0(g)$.
(b) If $a \neq 0$, then $S_a : {}^A K \times K \to K$, and if $h | A - \{a\} = g | A - \{a\}$, then $S_a(h, i) = S_a(g, i)$ for $i = 0, 1$.
(c) If $a \neq 0$, then for every $h \in {}^A K$, we have $S_a(h, S_0(h)) = h(a)$.
(d) If $a \neq 0$, then for every $h \in {}^A K$, we have $S_a(h, 0) \neq S_a(h, 1)$.

To see that (d) is true, assume for contradiction that for some $h \in {}^A K$ and some $a \neq 0$, we have $S_a(h, 0) = S_a(h, 1) = $ (say) 0. It now follows that $S_a(h, S_0(h)) = 0$. Choose $g \in {}^A K$ so that $g | A - \{a\} = h | A - \{a\}$, but $g(a) \neq h(a)$. By (b), we know that $S_a(g, 0) = S_a(h, 0) = 0$ and $S_a(g, 1) = S_a(h, 1) = 0$. Thus, $S_a(g, S_0(g)) = 0$. But then we have $g(a) = S_a(g, S_0(g)) = 0 = S_a(h, S_0(h)) = h(a)$, which is a contradiction. This proves (d).

Define $P = \langle G_a : a \in A \rangle$ as follows: Given $h \in {}^A K$, we set $G_0(h) = S_0(h)$, and for $a \neq 0$, we set $G_a(h) = S_a(h, h(0))$. That is, agent a is guessing as if agent 0 had signaled with agent 0's actual hat color. Assume now that $h \in {}^A K$. We claim that with P, either everyone guesses correctly or everyone guesses incorrectly.

Case 1: $S_0(h) = h(0)$.
    In this case, $G_0(h) = S_0(h) = h(0)$, and if $a \neq 0$, then $G_a(h) = S_a(h, h(0)) = S_a(h, S_0(h)) = h(a)$. Hence, in this case, everyone guesses correctly using P.
Case 2: $S_0(h) \neq h(0)$.
    In this case, $G_0(h) = S_0(h) \neq h(0)$, and if $a \neq 0$, then $G_a(h) = S_a(h, h(0)) \neq S_a(h, S_0(h)) = h(a)$, where we have used (d) above to conclude that $S_a(h, h(0)) \neq S_a(h, S_0(h))$. Hence, in this case, everyone guesses incorrectly using P.

Assume now that $K = 2$, and that $P = \langle G_a : a \in A \rangle$ is a predictor as in (2). As pointed out in Sect. 3.3, the assumption that everyone can see everyone else's hat in Theorem 3.3.1 is necessary. Let $S = \langle S_a : a \in A \rangle$ be the following signaling strategy: Given $h \in {}^A K$, we set $S_0(h) = G_0(h)$, and for $a \neq 0$, we set $S_a(h, S_0(h)) = G_a(h)$ iff $S_0(h) = h(0)$. We claim that if $a \neq 0$, then $S_a(h, S_0(h)) = h(a)$.

Case 1: $S_0(h) = h(0)$.

Because $S_0(h) = G_0(h)$, we know $G_0(h) = h(0)$ in this case, and so everyone must guess correctly given h and using P. Thus, $S_a(h, S_0(h)) = G_a(h) = h(a)$, as desired.

Case 2: $S_0(h) \neq h(0)$.

Because $S_0(h) = G_0(h)$, we know $G_0(h) \neq h(0)$ in this case, and so everyone must guess incorrectly given h and using P. Thus, $S_a(h, S_0(h)) \neq G_a(h) \neq h(a)$, so $S_a(h, S_0(h)) = h(a)$, as desired.   □

For Gabay's extension of the hat problem in Sect. 2.4, suppose there are countably many agents labeled by natural numbers and lined up in order so that everyone sees the hats of higher-numbered agents, and hears the guesses of lower-numbered agents. To obtain a strategy under which everyone but possibly agent 0 guesses correctly, one simply uses Lenstra's theorem and (2) implies (1) in Theorem 3.3.2, together with the fact that hearing correct answers from agents less than m is, for agent m, the same as being able to see everyone but agent 0.

3.4   The Role of the Axiom of Choice

In this section (largely drawn from [HT08a] and [HT10]) we show that the system ZF + DC is not strong enough to prove Lenstra's theorem or the Gabay-O'Connor theorem, even when restricted to the case of two colors and countably many agents. Historically, the precursor to our results here is a slightly weaker observation (in a different but related context) of Roy O. Davies that was announced in [Sil66].

Let BP be the assertion that every set of reals has the property of Baire.

It is known (assuming ZF is consistent) that ZF + DC cannot disprove BP [JS93]. (This was established earlier, assuming the existence of a large cardinal, in [Sol70].) It follows that ZF + DC cannot prove any theorem that contradicts BP, as any such proof could be turned into a proof that BP fails. We will show that Lenstra's theorem and the Gabay-O'Connor theorem contradict BP, and thus ZF + DC cannot prove Lenstra's theorem or the Gabay-O'Connor theorem. While BP is very useful for establishing results such as these, one should note that BP is false in ZFC (for instance, ZFC can prove Lenstra's theorem, which contradicts BP).

Let T_k be the measure-preserving homeomorphism from $^\omega 2$ to itself that toggles the kth bit in a sequence of 0s and 1s. Call a set $D \subseteq {}^\omega 2$ a *toggle set* if there are infinitely many values of k for which $T_k(D) \cap D = \emptyset$.

The next lemma is key to the results in this section; its proof makes use of the following observation. If a set D has the property of Baire but is not meager, then there exists a nonempty open set V such that the symmetric difference of D and V is meager. Hence, if we take any basic open set $[s] \subseteq V$, then it follows that $[s] - D$ is meager.

Lemma 3.4.1. *Every toggle set with the property of Baire is meager.*

Proof. Assume for contradiction that D is a non-meager toggle set with the property of Baire, and choose a basic open set $[s]$ such that $[s] - D$ is meager. Because D is a toggle set, we can choose k greater than the length of s such that $T_k(D) \cap D = \emptyset$. It now follows that $[s] \cap D \subseteq [s] - T_k(D)$. But $T_k([s]) = [s]$, because k is greater than the length of s. Hence, $[s] \cap D \subseteq [s] - T_k(D) = T_k([s]) - T_k(D) = T_k([s] - D)$. Thus, $[s] \cap D$ is meager, as was $[s] - D$. This means that $[s]$ itself is meager, a contradiction. □

With these preliminaries, the following theorem (of ZF + DC) shows that Lenstra's theorem contradicts BP, and hence it cannot be proven without some nontrivial version of the axiom of choice.

Theorem 3.4.2. *Consider the situation in which the set A of agents is countably infinite, there are two colors, and each agent sees all of the other hats. Assume BP. Then for every predictor there exists a coloring under which someone guesses correctly and someone guesses incorrectly.*

Proof. Assume that P is a predictor and let D denote the set of colorings for which P yields all correct guesses, and let I denote the set of colorings for which P yields all incorrect guesses. Notice that both D and I are toggle sets, since changing the hat on one agent causes his (unchanged) guess to switch from right to wrong or vice versa. If D and I both have the property of Baire, then both are meager. Choose $h \in {}^\omega 2 - (D \cup I)$. Under h, someone guesses correctly and someone guesses incorrectly. □

In ZFC, non-meager toggle sets do exist: as seen in the above proof, if all toggle sets are meager, then Lenstra's theorem fails, but Lenstra's theorem is valid in ZFC.

We derived Lenstra's theorem from the Gabay-O'Connor theorem, so Theorem 3.4.2 also shows us that some nontrivial version of the axiom of choice is needed to prove the Gabay-O'Connor theorem. However, the Gabay-O'Connor theorem, even in the case of two colors and countably many agents, is stronger than the assertion that the corresponding hat problem has a solution: the theorem does not require that agents can see *all* other hats, and it produces not just a predictor, but a robust predictor. The following theorem (of ZF + DC) shows that any solution to the Gabay-O'Connor hat problem, even in the countable case, contradicts BP and hence requires some nontrivial version of the axiom of choice.

Theorem 3.4.3. *Consider the case of the Gabay-O'Connor hat problem in which the set of agents is countably infinite. Assume BP. Then for every predictor there exists a coloring under which the number of agents guessing incorrectly is infinite.*

Proof. Assume that P is a predictor and, for each k, let D_k denote the set of colorings for which P yields all correct guesses from agents numbered k and higher. Notice that each D_k is a toggle set, since changing the hat on an agent higher than k causes his (unchanged) guess to switch from right to wrong. If all the D_ks have the property of Baire, then all are meager. Let D be the union of the D_ks, and choose $h \in {}^\omega 2 - D$. Under h, the number of agents guessing incorrectly is infinite. □

Theorems 3.4.2 and 3.4.3 can be recast in the context of Lebesgue measurability, to show that Lenstra's theorem and the Gabay-O'Connor theorem both imply the existence of non-measurable sets of reals. However, to show that ZF + DC cannot prove the existence of non-measurable sets of reals, one must assume the consistency of ZFC plus the existence of a large cardinal [Sol70, She84]. Although this is not a particularly onerous assumption, it is why we favored the presentation in terms of the property of Baire.

3.5   The Role of Square Bracket Partition Relations

As we said earlier, as long as the set of colors is finite there are trivial predictors ensuring that infinitely many agents guess correctly. Thus, the startling part of the predictor in the Gabay-O'Connor theorem is that "almost everyone" guesses correctly when using them. Intuitively, it is tempting to link the need for the axiom of choice with the goal of having a finite-error predictor.

This intuition, however, is somewhat incomplete. It turns out (as we momentarily show) that if the set of colors is infinite, then some nontrivial version of the axiom of choice is needed to obtain a predictor ensuring that *at least one* agent guesses correctly, even if every agent can see every other agent. To prove this, we need a preliminary definition that explains the "arrow notation" used in discussing generalizations of Ramsey's theorem.

Definition 3.5.1. The notation $\omega \to (\omega)_2^\omega$ means that for every function $f : [\omega]^\omega \to 2$, there exists an infinite set X such that f is constant on $[X]^\omega$. Similarly, the notation $\omega \to [\omega]_\omega^\omega$ means that for every function $f : [\omega]^\omega \to \omega$, there exists an infinite set X and a number $n \in \omega$ such that $n \notin f([X]^\omega)$.

Adrian Mathias [Mat77] showed that if ZFC plus the existence of a large cardinal is consistent, then so is ZF + DC together with the assertion that $\omega \to (\omega)_2^\omega$. We only need the (apparently) weaker assertion $\omega \to [\omega]_\omega^\omega$, and the following easy consequence of it.

Lemma 3.5.2. *Assume* $\omega \to [\omega]_\omega^\omega$. *Then for every* $X \in [\omega]^\omega$ *and every function* $f : [X]^\omega \to \omega$, *there exists an infinite set* $X' \subseteq X$ *and a number* $n \in X$ *such that* $n < \min(X')$ *and for every* $Y \in [X']^\omega$, *we have* $f(Y) \neq n$.

Proof. Given $f : [X]^\omega \to \omega$, define $g : [X]^\omega \to X$ by setting $g(Y) = f(Y)$ if $f(Y) \in X$ and $g(Y) = \min(X)$ otherwise. Because $\omega \to [\omega]_\omega^\omega$, we can (by identifying X with ω) choose $X'' \in [X]^\omega$ and $n \in X$ such that $n \notin f([X'']^\omega)$. Letting $X' = X'' - \{0, \ldots, n\}$ works. $\square$

The following theorem is inspired by results in Fred Galvin and Karel Prikry's 1976 paper [GP76].

Theorem 3.5.3. *Consider the version of the hat problem in which $A = \omega$, $K = \omega$, and each agent gets to see all of the other hats. Assume that $\omega \rightarrow [\omega]^\omega_\omega$. Then for every predictor there exists a hat coloring under which every agent guesses incorrectly.*

Proof. Assume that $\omega \rightarrow [\omega]^\omega_\omega$ and that $P = \langle G_n : n \in \omega \rangle$ is a predictor in which everyone gets to see everyone else's hat. We will inductively construct a sequence $\langle (x_n, X_n) : n \in \omega \rangle$ of pairs such that $x_0 < x_1 < \cdots$ and $X_0 \supset X_1 \supset \cdots$ and such that for each $n \in \omega$, the following hold:

(a) $x_n \in X_n$ and $X_n \in [\omega]^\omega$.
(b) $x_n < \min(X_{n+1})$.
(c) For each $Y \in [X_{n+1}]^\omega$, if $\langle y_0, y_1, \ldots \rangle$ is the increasing enumeration of Y, then $G_n(\langle x_0, \ldots, x_{n-1}, *, y_0, y_1, \ldots \rangle) \neq x_n$ for any value of $*$. Notice that $G_n(\langle x_0, \ldots, x_{n-1}, *, y_0, y_1, \ldots \rangle)$ is uniquely determined regardless of what number is substituted for $*$ (i.e., agent n cannot see his own hat).

We begin with $X_0 = \omega$. Suppose now that $n \geq 0$ and that we have constructed X_n as well as x_k for each k with $0 \leq k < n$. Let $f : [X_n]^\omega \rightarrow \omega$ be given by $f(Y) = G_n(\langle x_0, \ldots, x_{n-1}, *, y_0, y_1, \ldots \rangle)$. By Lemma 3.5.2, we can choose $x_n \in X_n$ and $X_{n+1} \in [X_n]^\omega$ such that $x_n < \min(X_{n+1})$ and for each $Y \in [X_{n+1}]^\omega$, $f(Y) \neq x_n$. That is, for each $Y \in [X_{n+1}]^\omega$, if $\langle y_0, y_1, \ldots \rangle$ is the increasing enumeration of Y, then $G_n(\langle x_0, \ldots, x_{n-1}, *, y_0, y_1, \ldots \rangle) \neq x_n$. Thus the coloring $\langle x_0, x_1, \ldots \rangle$ will make every agent guess incorrectly. □

3.6 Open Questions

The single most prominent open question related to the material in this chapter is the following:

Question 3.6.1. Is the Gabay-O'Connor theorem equivalent to the axiom of choice?

We also don't know the answer to the following question about the partition relations that we used in this chapter.

Question 3.6.2. Does $\omega \rightarrow [\omega]^\omega_\omega$ imply $\omega \rightarrow (\omega)^\omega_2$?

Question 3.6.3. Assume ZF + DC + the assumption that for every predictor with $A = \omega$ and $K = \omega$ there exists a hat coloring under which every agent guesses incorrectly. Does $\omega \rightarrow [\omega]^\omega_\omega$? Does $\omega \rightarrow (\omega)^\omega_2$?

Question 3.6.4. Assuming ZF + DC + $^\omega 2$ is compact, Proposition 1.3.7 showed that if $V(n)$ is finite for each n, then for every predictor there is a hat coloring making everyone guess incorrectly. Assume ZF + DC + the assumption that if $V(n)$ is finite for each n, then for every predictor there is a hat coloring making everyone guess incorrectly. Can one prove $^\omega 2$ is compact?

Chapter 4
The Denumerable Setting: One-Way Visibility

4.1 Background

In this chapter, we again have ω as the set of agents, but we only consider *one-way visibility* on ω. That is, if x sees y then $x < y$. Under this constraint, there is an obvious one-to-one correspondence between undirected graphs and one-way visibility graphs on ω, by saying that a sees b iff $a < b$ and a is adjacent to b in the undirected graph. We shall conflate the two notions and speak simply of *graphs* on ω, with the following conventions to avoid ambiguity: we use $V(a)$ to denote the set of agents that a can see, as opposed to those connected to a by an edge; aVb means that $a < b$ and a is adjacent to b in V; V is *transitive* if aVb and bVc implies aVc. We need one preliminary proposition regarding one-way visibility.

Proposition 4.1.1. *If V is a graph on ω, then for the hat problem with one-way visibility given by V (and any number of colors), there is a predictor ensuring one correct guess iff there is a predictor ensuring infinitely many correct guesses.*

Proof. Suppose that P is a predictor for which there is a coloring f such that the set of agents who guess correctly for f with P is finite. We will produce a coloring f' such that no agent guesses correctly for f' with P. Choose $n \in \omega$ such that no agent $a \geq n$ guesses correctly for f using P. We will construct a sequence $\langle f_0, \ldots, f_n \rangle$ of colorings such that no agent $a \geq n - i$ guesses correctly for f_i using P. We start with $f_0 = f$. Suppose now that $1 \leq i \leq n$ and that f_{i-1} has been defined. If agent $n - i$ guesses incorrectly for f_{i-1} using P, we let $f_i = f_{i-1}$. Otherwise, we change agent $n - i$'s hat, and let f_i be this new coloring. With the sequence of colorings constructed, we can let $f' = f_n$ and have the desired coloring f' for which no agent guesses correctly. $\square$

Our starting point in Sect. 4.2 is to use the μ-predictor from Chap. 1 to completely characterize those *transitive* graphs adequate for minimal and optimal predictors. In Sect. 4.3, we handle the nontransitive characterization for optimal predictors. For minimal predictors, this is still an open question, and we explore a particular

C.S. Hardin and A.D. Taylor, *The Mathematics of Coordinated Inference*,
Developments in Mathematics 33, DOI 10.1007/978-3-319-01333-6_4,
© Springer International Publishing Switzerland 2013

visibility graph called the "parity relation" in Sects. 4.4 and 4.5 (both sections are based on work in [HT10]).

Section 4.6 (from [T12]) deals with P-point ultrafilters and Ramsey ultrafilters on ω, and this requires a few definitions that we now motivate and present. Section 4.7 provides a finer analysis of the investigations begun in Sect. 4.6, and Sect. 4.8 provides a brief glimpse at the work done by Andreas Blass and others on evasion and prediction.

On ω, a minimal predictor will be one ensuring at least one correct guess (although, by Proposition 4.1.1, this is equivalent to ensuring infinitely many correct guesses). So if $I = [\omega]^{<\omega}$, then a minimal predictor is one ensuring that the set of agents guessing correctly is of positive I-measure and a finite-error predictor is one ensuring that the set of agents guessing correctly is of I-measure one.

In the context of ideals, it is often more natural to use terminology based on the set of agents who guess correctly than on the set of agents who guess incorrectly. Hence, with $I = [\omega]^{<\omega}$, we will sometimes speak of positive I-measure predictors rather than minimal predictors and I-measure one predictors rather than finite-error predictors; we also use similar terminology with other ideals and ultrafilters.

4.2   Optimal and Minimal Predictors for Transitive Graphs

In this section, we consider two questions: For which transitive visibility graphs on ω do we get a finite-error predictor? For which transitive visibility graphs on ω do we get a minimal predictor?

Theorem 4.2.1. *Suppose V is a graph on ω and consider the hat problem with one-way visibility given by V. Assume also that V is transitive. Then the following are equivalent:*

1. *The graph V contains no infinite independent subgraph.*
2. *There exists a finite-error predictor for any set K of colors.*
3. *There exists a finite-error predictor for two colors.*

Proof. Suppose first that V contains no infinite independent set, and for each $a \in \omega$, we have $\equiv_a$ defined on $^\omega K$ by $f \equiv_a g$ iff $f(b) = g(b)$ for every b that a can see. We claim that the μ-predictor $M_\prec$ from Sect. 1.4 is the desired finite-error predictor. (Recall that $M_\prec(f)(a) = \langle f \rangle_a(a)$ for every $a \in A$ where $\prec$ is a fixed well-ordering of the colorings and $\langle f \rangle_a$ is the $\prec$-least coloring that agent a cannot distinguish from the coloring f.)

Suppose for contradiction that there are infinitely many agents who guess incorrectly for some coloring f while using the μ-predictor. By Ramsey's theorem and the fact that V contains no infinite independent set, we can assume that, among these agents guessing incorrectly, i can see j whenever $i < j$. The following two claims now yield an infinite descending chain in the well-ordering of $^\omega K$, and this will be our desired contradiction.

Claim 1. If $i < j$, then $\langle f \rangle_i \succeq \langle f \rangle_j$.

Proof. Because $i < j$ we have $V(i) \supseteq V(j)$ by transitivity. Hence $[f]_i \subseteq [f]_j$, because any coloring consistent with what i sees will be consistent with what j sees. Hence the $\prec$-least element $\langle f \rangle_j$ of the bigger set $[f]_j$ will be at least as small as the $\prec$-least element $\langle f \rangle_i$ of the smaller set $[f]_i$.

Claim 2. If $i < j$, then $\langle f \rangle_i \neq \langle f \rangle_j$.

Proof. Because i sees j, we have that $\langle f \rangle_i (j) = f(j)$. But because j guesses incorrectly for the coloring f, we have $\langle f \rangle_j (j) \neq f(j)$. Thus $\langle f \rangle_i (j) \neq \langle f \rangle_j (j)$ so $\langle f \rangle_i \neq \langle f \rangle_j$.

This completes the proof that (1) implies (2). The proof that (2) implies (3) is trivial. Finally, for (3) implies (1) we invoke Corollary 1.3.3 which guarantees that for any predictor, we can make everyone in an independent set wrong. □

Turning now to minimal strategies, there is a very satisfying characterization of the amount of visibility needed in the case where the visibility graph V is transitive.

Theorem 4.2.2. *Suppose V is a graph on ω and we consider the hat problem with one-way visibility given by V. Assume also that V is transitive. Then the following are equivalent:*

1. *The graph V contains an infinite path $x_0 V x_1 V x_2 V \cdots$.*
2. *The graph V contains an infinite complete subgraph.*
3. *For any set K of colors, there is a predictor ensuring infinitely many correct guesses.*
4. *For two colors, there is a predictor ensuring infinitely many correct guesses.*
5. *For two colors, there is a predictor ensuring at least one correct guess.*

Proof. The fact that (1) implies (2) is immediate, because V is transitive, and (2) implies (3) because we can use the μ-predictor on the agents in the complete subgraph (and simply ignore all other agents). As (3) implies (4) and (4) implies (5) are also trivial, it suffices to prove that (5) implies (1).

Assume that V contains no infinite path and let P be any predictor for two colors. We will use Proposition 1.3.2 to produce a coloring that makes everyone guess wrong, and for this it suffices to show that $B(\omega) = \omega$. To see this, suppose not and note that if $x \notin B(\omega)$, then $V(x) \nsubseteq B(\omega)$. Using this we can immediately construct an infinite path $x_0 V x_1 V x_2 V \cdots$, which we are assuming does not exist. □

Theorems 4.2.1 and 4.2.2 are generalized to the context of ideals in Sect. 5.3.

4.3   Characterizing Graphs Yielding Finite-Error Predictors

We now generalize Theorem 4.2.1 to the case of graphs on ω that are not necessarily transitive. This result is due to the first author and can be found in [H10].

Theorem 4.3.1. *Suppose V is a graph on ω and we consider the hat problem with one-way visibility given by V. Then the following are equivalent:*

1. *The graph V contains no infinite independent subgraph.*
2. *There exists a finite-error predictor for any set K of colors.*
3. *There exists a finite-error predictor for two colors.*

Proof. To prove that (1) implies (2), we will inductively throw edges in the graph V away until we arrive at a subgraph V' (still with vertex set ω) that is transitive. Moreover, we will do this in such a way that V' still contains no infinite independent set. At this point the proof that (1) implies (2) follows immediately from Theorem 4.2.1, because we can just ignore visibilities corresponding to the edges that we throw out in arriving at V'.

At stage n, the only edges that will be deleted from V will have right endpoint n. Stage n itself will consist of n steps, where at step j we make a decision as to whether V' will have an edge from j to n. If there is no edge from j to n in V, then there will certainly be no edge from j to n in V', as we want V' to be a subgraph of V. However, if we have jVn in V, then we will delete this edge if there is some $i < j$ such that, in the subgraph so far constructed, we have $iVjVn$ but i not adjacent to n. If there is no such $i < n$, we leave the edge from j to n in and move on to step $j + 1$.

Claim 1. V' is transitive.

Proof. Suppose that we have $i < j < n$ and, in V', $iVjVn$. Then we must also have iVn or else at step j of stage n we'd have thrown out the edge from j to n.

Claim 2. V' contains no infinite independent set.

Proof. Assume that V' contains an infinite independent set. We first construct from this assumption what might be called a "leftmost" infinite independent set, and then we'll use this to show that there must have already existed an infinite independent set in V.

Choose an infinite independent set with smallest possible first element, and call this element x_0. Now, among all infinite independent sets with first element x_0, choose one with smallest possible second element and call this element x_1. Continuing this yields the independent set we want as $\{x_0, x_1, \ldots\}$.

We now claim that each x_j is adjacent to only finitely many x_n in the original graph V; this will give us our desired contradiction as it immediately yields an infinite independent set in V.

So suppose for contradiction that we have x_j adjacent to infinitely many x_n in V. Then, in passing to V', all these edges had to be thrown away. Each such deleted edge was based on some $i < x_j$, and so there must be a single such i corresponding to infinitely many of the x_n. In particular, this means that iVx_j (so $i \neq x_p$ for any $p < n$) and i is independent from x_n for infinitely many n.

We can't have $i < x_0$ because i, together with the infinitely many x_n that it is independent from would contradict our choice of x_0. Let x_p be the largest of the original x_ns that is less than i. Then x_p, i, and the x_ns that are independent from i contradict our choice of x_{p+1}.

This completes the proof that (1) implies (2). The proofs that (2) implies (3) and (3) implies (1) are exactly as in Theorem 4.2.1.   □

In the next section we show that the result for minimal predictors in the transitive setting does not hold in the nontransitive setting.

4.4 ZFC Results for the Parity Relation

For some kinds of hat problems, the results in the transitive case carry over to the nontransitive case, but are just (apparently) harder to prove. For example, in the transitive case with two or more colors, a finite-error predictor exists iff there is no sequence x_n of agents such that x_i cannot see x_j for $i \leq j$, and the proof is fairly succinct; in the nontransitive case, the same holds when there are countably many agents (the uncountable case is open), but the proof, as seen in the last section, is more elaborate. So, it would be natural to speculate that the situation with minimal predictors is similar, and that Theorem 4.2.2 holds in the nontransitive case, but with a more complicated proof. However, it turns out that the nontransitive case is very unlike the transitive case here, and that minimal predictors can exist without having an infinite chain of visibility. Moreover, the existence of minimal predictors can now depend on the number of colors, in some cases yielding independence results.

To exhibit these phenomena that distinguish the nontransitive case from the transitive case, we consider in this section and the next two the visibility graph on ω in which evens see higher-numbered odds and odds see higher-numbered evens; we call this the *parity relation*, and denote the graph by EO. This directed graph is (essentially) isomorphic to the one on $A = 2 \times \omega$ where (i, m) sees (j, n) iff $i \neq j$ and $m < n$. In what follows, we will freely switch back and forth between ω and $2 \times \omega$.

The parity relation is notable for having many infinite paths but being highly nontransitive in the sense that when $x\mathrm{EO}y$ and $y\mathrm{EO}z$ we *never* have $x\mathrm{EO}z$.

Notationally, we will use A^0 for both the set of evens and for $\{0\} \times \omega$ and A^1 for both the set of odds and for $\{1\} \times \omega$. Similarly, we will often think of a hat assignment f as a pair (f^0, f^1) where f^i specifies the hat assignment for A^i, and a predictor P as a pair (P^0, P^1) where P^i tells us how the agents in A^i guess (as a function of f^{1-i}).

Our starting point for discussing EO is the following theorem; the $n = 2$ case appeared in [HT10] and the general case was obtained by D. J. Velleman [Vel11]. The proof relies on Theorem 2.2.3.

Theorem 4.4.1. *If there are finitely many colors, then there is a predictor for the parity relation* EO *that ensures at least one correct guess.*

Proof. Assume there are $k > 1$ colors, and let n be the number of agents in team 1 in the proof of Theorem 2.2.3. Split the odd agents into infinitely many teams of size n, and the even agents into infinitely many teams of size $k - 1$. Pair each even team of size $k - 1$ with an odd team of size n with higher numbers. Each even team will proceed as in the proof of Theorem 2.2.3 with the odd team it has been paired with. Since the odd team has higher numbers, the even team can see their hats, so they can play the strategy from Theorem 2.2.3. Unfortunately, the odd team can't see the hats of the even team. So they will guess the colors of the hats of the even team and then play according to this guess. If any odd team guesses correctly, then someone will guess his hat color correctly.

Since there are infinitely many teams of size $k - 1$ among the even agents, there is at least one i such that infinitely many even teams have hat coloring h_i. Let i_0 be the least such i. Since each odd agent can see all but finitely many of the even hat colors, the odd agents all know i_0. They all guess that their opposing team has hat coloring h_{i_0} and play accordingly. Then for each of the infinitely many even teams that has hat coloring h_{i_0}, the odd team that they are paired with correctly guesses their hat coloring and therefore plays correctly according to the strategy in the proof of Theorem 2.2.3. Therefore someone guesses his hat color correctly. $\qquad\square$

At the other extreme (in terms of number of colors) we have the following.

Theorem 4.4.2. *If there are ω_2 colors, then there is no predictor for the parity relation EO that will ensure even one correct guess.*

Proof. Fix any predictor $P = (P^0, P^1)$. For any ordinal α, let c_α be the function on ω that is constantly α. We intend to color A^0 with c_β for some $\beta \in \omega_1$, and color A^1 with c_γ for some $\gamma \in \omega_2$.

Since ω_2 is regular, we can choose $\gamma \in \omega_2$ such that $\gamma > P^1(c_\alpha)(n)$ for every $\alpha \in \omega_1$ and $n \in \omega$. Now choose $\beta \in \omega_1$ such that $\beta > P^0(c_\gamma)(n)$ for every $n \in \omega$ such that $P^0(c_\gamma)(n) \in \omega_1$. Then, under the hat assignment (c_β, c_γ), everyone guesses incorrectly. $\qquad\square$

What is really going on in the above argument is that given functions $S^0 : \omega_2 \to \omega_1$ and $S^1 : \omega_1 \to \omega_2$, there is a pair (β, γ) such that $\beta > S^0(\gamma)$ and $\gamma > S^1(\beta)$: just choose γ above the supremum of S^1, and β above $S^0(\gamma)$.

4.5 Independence Results for the Parity Relation

Somewhat surprisingly, there seems to be a close connection between the existence of minimal predictors for the parity relation EO and properties of the ideal of meager sets of real numbers. In fact, our consistency results make use of two prominent cardinal invariants; these and several others all lie between $\aleph_1$ and $2^{\aleph_0}$, and the relationships between them are well understood; a summary can be found in [Jec03, pp. 532–533] while [BJ95, Bar10, Bla96, Mil81] provide detailed accounts. We say that functions f and g *infinitely agree* if $f \cap g$ is infinite, and *only finitely agree* if $f \cap g$ is finite.

Definition 4.5.1. With $\mathcal{M}$ denoting the ideal of meager subsets of $\mathbf{R}$,

(a) $\text{cov}(\mathcal{M})$ is the least cardinality of a subset of $\mathcal{M}$ whose union is $\mathbf{R}$;
(b) $\text{non}(\mathcal{M})$ is the least cardinality of a nonmeager set.

Lemma 4.5.2 ([BJ95, pp. 54–59]).

(a) $\text{cov}(\mathcal{M})$ *is the smallest size of a family* $F \subseteq {}^{\omega}\omega$ *such that* $(\forall g \in {}^{\omega}\omega)(\exists f \in F)$ f *and* g *only finitely agree.*
(b) $\text{non}(\mathcal{M})$ *is the smallest size of a family* $F \subseteq {}^{\omega}\omega$ *such that* $(\forall g \in {}^{\omega}\omega)(\exists f \in F)$ f *and* g *infinitely agree.*

The following definition and lemma are key to the positive results for the parity relation EO.

Definition 4.5.3. A family G is *agreeable* if for any $F \subseteq G$ with $|F| < |G|$, there is a $g \in G$ such that for each $f \in F$, the functions g and f infinitely agree.

Lemma 4.5.4. *If* ${}^{\omega}\nu$ *is agreeable, then there is a predictor for the parity relation* EO *that ensures at least one correct guess.*

Proof. Let $\lambda = |{}^{\omega}\nu|$ and fix a well-ordering $\preceq$ of ${}^{\omega}\nu$ of order type λ. For any $f \in {}^{\omega}\nu$, let $\hat{f}$ be $\preceq$-minimal such that $\hat{f}$ and f eventually agree. The agents' strategies will be that for a hat assignment (f^0, f^1), the agents in A^i will assume $\hat{f}^i \preceq \hat{f}^{1-i}$ and guess according to a function that infinitely agrees with each $g \preceq \hat{f}^{1-i}$; at least one of those assumptions will turn out to be correct, yielding a minimal predictor.

We define $p : {}^{\omega}\nu \to {}^{\omega}\nu$ as follows. For any $f \in {}^{\omega}\nu$ we have $|\{g \in {}^{\omega}\nu : g \preceq \hat{f}\}| < \lambda$, so by the agreeability of ${}^{\omega}\nu$, we can choose $p(f)$ to infinitely agree with each $g \preceq \hat{f}$. Note that for a hat assignment (f^0, f^1), an agent in A^i can only see the values of f^{1-i} on a tail of ω, but this is enough information to determine $\hat{f}^{1-i}$ and hence $p(f^{1-i})$. The predictor $P = (P^0, P^1)$ is defined by letting $P^0 = P^1 = p$.

For any hat assignment (f^0, f^1), we have $\hat{f}^0 \preceq \hat{f}^1$ or $\hat{f}^1 \preceq \hat{f}^0$. Suppose the former. Then $p(f^1)$ infinitely agrees with $\hat{f}^0$, and since $\hat{f}^0$ and f^0 eventually agree, $p(f^1)$ also infinitely agrees with f^0, so infinitely many agents in A^0 guess correctly. Similarly, if $\hat{f}^1 \preceq \hat{f}^0$, infinitely many agents in A^1 guess correctly. $\square$

An immediate consequence of Lemma 4.5.4 is the following.

Theorem 4.5.5. *If* CH *holds and there are* $\aleph_1$ *colors, then there is a predictor for the parity relation* EO *that ensures at least one correct guess.*

Proof. It suffices to show that $^\omega\omega_1$ is agreeable. Note that under CH, $|^\omega\omega_1| = \aleph_1$. Take any $F \subseteq {}^\omega\omega_1$, $|F| < \aleph_1$. Since F is countable, we can produce a sequence of functions $f_0, f_1, \ldots \in {}^\omega\omega_1$ in which each $f \in F$ appears infinitely often. The function $g(n) = f_n(n)$ infinitely agrees with every $f \in F$. $\square$

In fact, in models of ZFC + non$(\mathcal{M}) = 2^{\aleph_0} = \aleph_2$, the existence of a minimal strategy with ω colors is equivalent to a number of natural conditions, as the following theorem shows.

Theorem 4.5.6. *Assume* non$(\mathcal{M}) = 2^{\aleph_0} = \aleph_2$. *Then the following are equivalent:*

1. $\operatorname{cov}(\mathcal{M}) = \aleph_2$.
2. MA$_{\aleph_1}$ (countable).
3. $^\omega\omega$ *is agreeable.*
4. *There is a predictor for the parity relation* EO *that ensures at least one correct guess with ω colors.*

Proof. (1)$\Leftrightarrow$(2) can be found in [BJ95, p. 138].

(1)$\Rightarrow$(3) is an easy consequence of Lemma 4.5.2(a), but we offer a direct argument of (2)$\Rightarrow$(3) for the sake of making the connection intuitively clear. Let Q be the partial order of finite partial functions from ω to ω, ordered by reverse inclusion, and note that Q is countable. Take any $F \subseteq {}^\omega\omega$ with $|F| < |^\omega\omega| = \aleph_2$. For $f \in F$ and $n \in \omega$, let $D_{f,n} = \{ q \in Q : n \in \operatorname{dom}(q) \wedge (\exists k \geq n)\, q(k) = f(k) \}$, which is dense in Q. Let $\mathcal{D} = \{ D_{f,n} : f \in F,\ n \in \omega \}$. We have $|\mathcal{D}| \leq \aleph_1$, so by MA$_{\aleph_1}$ (countable), there is a $\mathcal{D}$-generic filter $G \subseteq Q$. Letting $g = \cup Q \in {}^\omega\omega$, g infinitely agrees with each $f \in F$. Therefore, $^\omega\omega$ is agreeable.

(3)$\Rightarrow$(4) is immediate from Lemma 4.5.4.

(4)$\Rightarrow$(1): Supposing $\operatorname{cov}(\mathcal{M}) = \aleph_1$, we will show that there is no minimal strategy for ω colors. Let F be as in Lemma 4.5.2(a), with $|F| = \aleph_1$. Take any predictor $P = (P^0, P^1)$. We intend to color A^0 with some $f^0 \in F$. Let $F' = \{ P^1(f) : f \in F \}$. Since $|F'| \leq \aleph_1 < \operatorname{non}(\mathcal{M})$, Lemma 4.5.2(b) gives us $f^1 \in {}^\omega\omega$ such that $(\forall f \in F')$ f and f^1 only finitely agree. By our choice of F, there exists $f^0 \in F$ such that f^0 only finitely agrees with $P^0(f^1)$. Then P has only finitely many correct guesses under hat assignment $f = (f^0, f^1)$, so there is no predictor ensuring infinitely many correct guesses and thus, by Proposition 4.1.1, no predictor ensuring even one correct guess. $\square$

The above theorem establishes the independence of the existence of a minimal strategy for the parity relations with ω colors from ZFC + non$(\mathcal{M}) = 2^{\aleph_0} = \aleph_2$: adding $\aleph_2$ random reals to a model of CH yields a model in which $\operatorname{cov}(\mathcal{M}) = \aleph_1 + \operatorname{non}(\mathcal{M}) = 2^{\aleph_0} = \aleph_2$ [Mil81, p. 109], and it is well known that models of MA $+ 2^{\aleph_0} = \aleph_2$ have $\operatorname{cov}(\mathcal{M}) = \operatorname{non}(\mathcal{M}) = 2^{\aleph_0} = \aleph_2$.

4.6   The Role of P-Point and Ramsey Ultrafilters

In this section we consider the situation in which we have an ideal I on ω and we ask how much visibility is needed for an I-measure one predictor (which we refer to as an I^*-predictor); this material is taken from [T12]. The following theorem gives a natural sufficient condition and a natural necessary condition.

Theorem 4.6.1. *Suppose V is a graph on ω and consider the hat problem with one-way visibility given by V. Assume that I is an ideal on ω.*

(a) For there to exist an I^*-predictor, it is sufficient to have $V \in I^* \times I^*$.
(b) For there to exist an I^*-predictor, it is necessary to have no independent set in I^+.

Proof. Theorem 3.2.3 gives us (a). The proof of (b) consists of the observation that we can always make everyone in an independent set guess wrong by Corollary 1.3.3.   □

At this point, two questions suggest themselves. For which ideals is the obvious sufficient condition from (a) also necessary? And for which ideals is the obvious necessary condition from (b) also sufficient? We introduce some terminology for such ideals.

Definition 4.6.2. An *SIN ideal I on ω* is one for which the obvious sufficient condition for the existence of an I^*-predictor ($V \in I^* \times I^*$) is also necessary.

An *NIS ideal I on ω* is one for which the obvious necessary condition for the existence of an I^*-predictor (no independent set of positive I-measure) is also sufficient.

"SIN" stands for "sufficient is necessary" and "NIS" stands for "necessary is sufficient." We read each of these prefixes letter-by-letter ("S-I-N" instead of "sin") and thus use the article "an" as opposed to "a."

In Sect. 4.3 we showed that the ideal $I = [\omega]^{<\omega}$ is an NIS ideal (that is, for every graph on ω with no infinite independent set, there is a finite-error predictor). On the other hand, it is not hard to see (as we now demonstrate) that if I is an SIN ideal then I^* is an ultrafilter. To see this, suppose that I^* is not an ultrafilter and let $X \subseteq \omega$ be such that both X and $\omega - X$ are of positive I-measure. Consider the graph V in which there is an edge from x to y iff they are distinct points in X or distinct points in $\omega - X$. Then, for every n, we have $V(n) \notin I^*$ but the agents in X have a finite-error predictor among themselves as do the agents in $\omega - X$. Thus, there is a finite-error predictor and hence an I^*-predictor, showing that the sufficient condition is not necessary.

In part because of this observation, we restrict attention in this section to the question of when an ultrafilter $\mathcal{U}$ on ω is an SIN ultrafilter and when it is an NIS ultrafilter (although we return to the consideration of ideals in Sect. 5.3). Rather surprisingly, the answers turn out to involve two well-known and oft-studied classes.

Definition 4.6.3. An ultrafilter $\mathcal{U}$ on ω is a *P-point ultrafilter* if every function defined on a set in $\mathcal{U}$ is either constant on a set in $\mathcal{U}$ or finite-to-one on a set in $\mathcal{U}$. It is a *Ramsey ultrafilter* if every function defined on a set in $\mathcal{U}$ is either constant on a set in $\mathcal{U}$ or one-to-one on a set in $\mathcal{U}$.

It is known that CH (or MA) implies there are both Ramsey ultrafilters and P-point ultrafilters that are not Ramsey, but it is also consistent with ZFC that even P-point ultrafilters don't exist, a result of Saharon Shelah; see [Wim82]. Ramsey ultrafilters get their name from the fact that they are precisely the ones for which every graph on ω has either a complete subgraph in the ultrafilter or an independent subgraph in the ultrafilter. But our use for these classes is in the two upcoming theorems.

Theorem 4.6.4. *An ultrafilter on ω is an SIN ultrafilter iff it is a P-point ultrafilter.*

Proof. Suppose first that $\mathcal{U}$ is not a P-point and choose $f : X \to \omega$ such that $X \in \mathcal{U}$ and f is neither constant on a set in $\mathcal{U}$ nor finite-to-one on a set in $\mathcal{U}$. Let V be the graph on ω in which n is adjacent to m iff $n \neq m$ and $f(n) = f(m)$. No agent sees a set in $\mathcal{U}$. However, for each $k \in \omega$, the agents in $f^{-1}(\{k\})$ have a finite-error predictor among themselves, and the combined use of these predictors will ensure that the set of errors is a set upon which the function f is finite-to-one, and thus not in $\mathcal{U}$. Hence, we have a $\mathcal{U}$-predictor even though each agent sees only a set of agents of $\mathcal{U}$-measure zero.

Now suppose that $\mathcal{U}$ is a P-point and V is a graph on ω for which $\{n \in \omega : V(n) \in \mathcal{U}\} \notin \mathcal{U}$. Thus, the set $Y = \{n \in \omega : V(n) \notin \mathcal{U}\} \in \mathcal{U}$. Let $\langle S_n : n \in \omega\rangle$ be any predictor. We will produce a coloring for which a set in $\mathcal{U}$ of agents guesses incorrectly.

Let $X = \{n \in Y : \exists k \in Y, \{k, n\} \in V\}$. If $X \notin \mathcal{U}$ then $Y - X$ is an independent set in $\mathcal{U}$ and we can easily produce a coloring making everyone in this set guess incorrectly. If $X \in \mathcal{U}$, then consider the function f defined on X that maps n to the least $k < n$ such that $k \in Y$ and $\{k, n\} \in V$. Because $\mathcal{U}$ is a P-point, there is a set $Z \in \mathcal{U}$ such that f is either constant or finite-to-one on Z. If $f(Z) = \{k\}$, then $k \in Y$ and $V(k) \in \mathcal{U}$, contrary to the definition of Y. Thus, f is finite-to-one on Z.

Notice that each agent in Z sees only finitely many other agents in Z. It now follows from Proposition 1.3.7 that there is a coloring that makes everyone in Z guess incorrectly, and thus completes the proof.   □

Theorem 4.6.5. *An ultrafilter on ω is an NIS ultrafilter iff it is a Ramsey ultrafilter.*

Proof. Suppose first that $\mathcal{U}$ is a Ramsey ultrafilter and that V is a graph on ω having no independent subgraph with vertex set in $\mathcal{U}$. Because $\mathcal{U}$ is Ramsey it follows that there is a complete subgraph with vertex set $X \in \mathcal{U}$. But now we know that the agents in X have a finite-error strategy among themselves and this ensures correct guesses by a set of agents in $\mathcal{U}$ as desired.

Suppose now that $\mathcal{U}$ is not a Ramsey ultrafilter, and choose f defined on $X \in \mathcal{U}$ such that f is not constant on any set in $\mathcal{U}$ and f is not one-to-one on any set in $\mathcal{U}$. We consider two cases:

Case 1: There exists a set $Y \in \mathcal{U}$ such that f is finite-to-one on Y.

Let V be the graph on ω in which n is adjacent to m iff $n \neq m$ and $f(n) = f(m)$. If Z is an independent set in V, then $f|Z$ is one-to-one, and so $Z \notin \mathcal{U}$. Now as in the proof of the previous theorem, we have that each agent in Y sees only finitely many other agents in Y, and so for any predictor, Proposition 1.3.7 yields a hat coloring for which every agent in Y guesses incorrectly. This shows the necessary condition is not sufficient.

Case 2: f is not finite-to-one on any set in $\mathcal{U}$.

Let V be the graph on ω in which n is adjacent to m iff $n < m$ and $f(n) > f(m)$ (or vice-versa). Suppose $Z \in \mathcal{U}$; we'll show that Z is not an independent set in V. We know that f is not finite-to-one on Z so we can choose p such that infinitely many points of Z map to p. But $\{x \in Z : f(x) \leq p\} \notin \mathcal{U}$ and so we can choose $n \in Z$ such that $f(n) > p$. But now if we choose $m \in Z$ such that $m > n$ and $f(m) = p$, then $m, n \in Z$, $n < m$, and $f(n) > f(m)$ so we have an edge from V in Z. But for any predictor, we make everyone in the set X guess incorrectly by Proposition 1.3.2. This completes the proof.   □

We conclude this section with one more question whose answer turns out to involve P-point ultrafilters. Our starting point is the observation that for every ultrafilter $\mathcal{U}$ on ω and every graph V on ω, either V or V^c provides enough visibility for a $\mathcal{U}$-predictor, where V^c is the graph whose edge set is the complement of V's edge set. Is there an ultrafilter $\mathcal{U}$ for which there is a graph so that both it and its complement provide enough visibility for a $\mathcal{U}$-predictor? Are there other ultrafilters for which it is always exactly one of V and V^c that provides enough visibility? The following answers both of these questions.

Theorem 4.6.6. *For any ultrafilter $\mathcal{U}$ on ω, the following are equivalent:*

1. *$\mathcal{U}$ is a P-point ultrafilter.*
2. *$\mathcal{U}$ is an SIN ultrafilter.*
3. *The collection of graphs yielding a $\mathcal{U}$-predictor is closed under finite intersections.*
4. *There is no graph V for which V and V^c both yield a $\mathcal{U}$-predictor.*
5. *For every graph V, exactly one of V and V^c yields a $\mathcal{U}$-predictor.*

Proof. We know that (1) and (2) are equivalent. For (2) implies (3), notice that if $V_1, \ldots, V_n$ are graphs yielding $\mathcal{U}$-predictors and $X_k = \{x : V_k(x) \in \mathcal{U}\}$, then $X_k \in \mathcal{U}$ for each k. But then $X = X_1 \cap \cdots \cap X_k \in \mathcal{U}$ and for any $x \in X$ we have $V_1(x) \cap \cdots \cap V_n(x) \in \mathcal{U}$. This is sufficient to yield a $\mathcal{U}$-predictor for the intersection of the graphs. Now (3) implies (4) is trivial, as is (4) implies (5). Finally, for (5) implies (2), suppose that $\mathcal{U}$ is not an SIN-ultrafilter. Then there exists a graph V such that $\{n : V(n) \in \mathcal{U}\} \notin \mathcal{U}$, but for which there is a $\mathcal{U}$-predictor. But now V^c satisfies the sufficient condition for a $\mathcal{U}$-predictor, so we have both a graph and its complement yielding a $\mathcal{U}$-predictor.   □

4.7 $\mathcal{U}$-Predictors

We begin this section with a finer analysis of the relationship between an ultrafilter $\mathcal{U}$ on ω and a graph V on ω. Our starting point is with an obvious implication that holds for every ultrafilter $\mathcal{U}$ on ω and every graph V on ω.

$$\boxed{\text{For some } X \in \mathcal{U}, V|X \text{ is complete.}}$$

$$\Downarrow$$

$$\boxed{\text{For all } X \in \mathcal{U}, V|X \text{ is not independent.}}$$

An ultrafilter $\mathcal{U}$ is a Ramsey ultrafilter iff for every graph V there is a set $X \in \mathcal{U}$ such that either $V|X$ is complete or $V|X$ is independent. Thus, for a (fixed) ultrafilter $\mathcal{U}$, the implication above reverses (for every V) iff the ultrafilter is a Ramsey ultrafilter. We will denote this unique reversal for Ramsey ultrafilters with an implication sign (as above) followed by an "R" for "Ramsey" and an exclamation point for the uniqueness (i.e., the non-reversal for non-Ramsey ultrafilters). Our picture now becomes:

$$\boxed{\text{For some } X \in \mathcal{U}, V|X \text{ is complete.}}$$

$$\Downarrow \quad \Uparrow R!$$

$$\boxed{\text{For all } X \in \mathcal{U}, V|X \text{ is not independent.}}$$

For a first pass at a finer analysis of this relationship between ultrafilters and graphs, recall that $\mathcal{U}$ is a P-point if every function defined on a set in $\mathcal{U}$ is either constant on a set in $\mathcal{U}$ or finite-to-one on a set in $\mathcal{U}$. In a similar vein, an ultrafilter $\mathcal{U}$ is called a *Q-point* if every finite-to-one function defined on a set in $\mathcal{U}$ is one-to-one on a set in $\mathcal{U}$. It follows immediately from the definition of a Ramsey ultrafilter that $\mathcal{U}$ is Ramsey iff $\mathcal{U}$ is both a P-point and a Q-point.

If V is a graph on ω and $X \subseteq \omega$, then we'll say that $V|X$ is *complete (mod $[\omega]^{<\omega}$)* if $X - V(x)$ is finite for every $x \in X$. Similarly, we'll say that $V|X$ is *independent (mod $[\omega]^{<\omega}$)* if $X \cap V(x)$ is finite for every $x \in X$. Our picture now becomes the following.

$$\boxed{\text{For some } X \in \mathcal{U}, V|X \text{ is complete.}}$$

$$\Downarrow \quad \Uparrow Q!$$

$$\boxed{\text{For some } X \in \mathcal{U}, V|X \text{ is complete (mod } [\omega]^{<\omega}).}$$

$$\Downarrow \quad \Uparrow \; P!$$

For all $X \in \mathcal{U}$, $V|X$ is not independent (mod $[\omega]^{<\omega}$).

$$\Downarrow \quad \Uparrow \; Q!$$

For all $X \in \mathcal{U}$, $V|X$ is not independent.

As we've said, $\mathcal{U}$ is a Ramsey ultrafilter iff for every graph V there exists a set $X \in \mathcal{U}$ such that either $V|X$ is complete or $V^c|X$ is complete. The reversal of the implication for P-points is the assertion that $\mathcal{U}$ is a P-point iff for every graph V there exists a set $X \in \mathcal{U}$ such that either $V|X$ is complete (mod $[\omega]^{<\omega}$) or $V^c|X$ is complete (mod $[\omega]^{<\omega}$). There appear to be two reversals for Q-points, but one is just the contrapositive of the other applied to the complement of the graph. The underlying assertion is that $\mathcal{U}$ is a Q-point iff for every graph V, if there exists a set $X \in \mathcal{U}$ such that $V|X$ is complete (mod $[\omega]^{<\omega}$), then there exists a set $Y \in \mathcal{U}$ such that $V|Y$ is complete.

Turning now to $\mathcal{U}$-predictors, we also embellish our diagram of implications with a few more boxes, and these require some notation and terminology.

Definition 4.7.1. If V is a graph on ω, then a set $X \subseteq \omega$ will be called V-*scattered* if every nonempty subset Y of X contains a point y such that $V(y) \cap Y = \emptyset$.

The following definition is a slight modification of Definition 1.3.1.

Definition 4.7.2. If V is a graph on ω and $X \subseteq \omega$, then we define a sequence $\langle B_\alpha(X) : \alpha < \omega_1 \rangle$ by setting $B_0(X) = X$ and, for $\alpha > 0$, $y \in B_\alpha(X)$ iff $V(y) \subseteq \bigcup\{ B_\beta(X) : \beta < \alpha \}$. We let $B(X) = \bigcup\{B_\alpha(X) : \alpha < \omega_1\}$.

Recall that a predictor P is robust if it disregards finite differences; that is, if $f \Delta g$ is finite, then $P(f) = P(g)$.

Definition 4.7.3. A predictor P is $\mathcal{U}$-*robust* if, whenever $f \Delta g \notin \mathcal{U}$, $P(f) = P(g)$.

Our diagram now becomes the following:

1. For some $X \in \mathcal{U}$, $V|X$ is complete.

$$\Downarrow \quad \Uparrow \; Q!$$

2. For some $X \in \mathcal{U}$, $V|X$ is complete (mod $[\omega]^{<\omega}$).

$$\Downarrow \quad \Uparrow \; P!$$

3. $V \in \mathcal{U} \times \mathcal{U}$.

$$\Updownarrow$$

$$\boxed{\text{4. } V^c \notin \mathcal{U} \times \mathcal{U}.}$$

$$\Updownarrow$$

$$\boxed{\text{5. There exists a } \mathcal{U}\text{-robust } \mathcal{U}\text{-predictor for } V.}$$

$$\Downarrow \quad \Uparrow P!$$

$$\boxed{\text{6. There exists a robust } \mathcal{U}\text{-predictor for } V.}$$

$$\Downarrow \quad \Uparrow P$$

$$\boxed{\text{7. There exists a } \mathcal{U}\text{-predictor for } V.}$$

$$\Downarrow \quad \Uparrow P$$

$$\boxed{\text{8. If } X \notin \mathcal{U}, \text{ then } B(X) \notin \mathcal{U}.}$$

$$\Updownarrow$$

$$\boxed{\text{9. For all } X \in \mathcal{U}, X \text{ is not } V\text{-scattered.}}$$

$$\Updownarrow$$

$$\boxed{\text{10. For all } X \in \mathcal{U}, V|X \text{ has an infinite path.}}$$

$$\Downarrow \quad \Uparrow P!$$

$$\boxed{\text{11. For all } X \in \mathcal{U}, V|X \text{ is not independent (mod } [\omega]^{<\omega}).}$$

$$\Downarrow \quad \Uparrow Q!$$

$$\boxed{\text{12. For all } X \in \mathcal{U}, V|X \text{ is not independent.}}$$

Theorem 4.7.4. *The implications indicated in the above diagram are all valid.*

Proof. We begin with the proofs of the 11 implications from the top of the diagram to the bottom.

(1)$\Rightarrow$(2). If $V|X$ is complete, then $V|X$ is complete (mod $[\omega]^{<\omega}$).

(2)$\Rightarrow$(3). If $V|X$ is complete (mod $[\omega]^{<\omega}$), then $V(x)$ is cofinite in X for each $x \in X$, and thus $V(x) \in \mathcal{U}$ for each $x \in X$. Hence, $V \in \mathcal{U} \times \mathcal{U}$.

(3)$\Leftrightarrow$(4). $\mathcal{U} \times \mathcal{U}$ is an ultrafilter, so $V \in \mathcal{U} \times \mathcal{U}$ iff $V^c \notin \mathcal{U} \times \mathcal{U}$.

(3)$\Rightarrow$(5) is a special case of Theorem 3.2.3, the more general form of the Gabay-O'Connor theorem.

(5)$\Rightarrow$(6)$\Rightarrow$(7) are trivial.

(7)$\Rightarrow$(8). Assume that $B(X) \in \mathcal{U}$ for some $X \notin \mathcal{U}$. Then, for any predictor, we can produce a hat coloring for which all the agents in $B(X) - X$ guess incorrectly, noting that $B(X) - X \in \mathcal{U}$. To do this, we begin by placing hats on the agents in X arbitrarily and then proceed inductively on the sets $B_\alpha(X)$.

(8)$\Rightarrow$(9). Suppose we have a set $Y \in \mathcal{U}$ that is V-scattered. Let $X = Y^c$. Then $X \notin \mathcal{U}$, but we claim that $B(X) = \omega$, and is thus in $\mathcal{U}$. To see this, assume that $\omega - B(X) \neq \emptyset$, and note that $\omega - B(X) \subseteq Y$. Choose $y \in \omega - B(X)$ such that $V(y) \cap (\omega - B(X)) = \emptyset$, and thus $V(y) \subseteq B(X)$. It now follows that $y \in B(X)$, and this is a contradiction.

(9)$\Rightarrow$(10). Assume $X \in \mathcal{U}$. Then X is not V-scattered so there exists a nonempty $Y \subseteq X$ such that for every $y \in Y$, we have $V(y) \cap Y \neq \emptyset$. Thus having chosen $y_0, \ldots, y_n \in Y$, we can choose $y_{n+1} \in V(y_n) \cap Y$. This yields the desired infinite path in X.

(10)$\Rightarrow$(11). Assume there exists a set $X \in \mathcal{U}$ that is independent (mod $[\omega]^{<\omega}$). We define a sequence $\langle B_n : n \in \omega \rangle$ of consecutive blocks of integers in X by choosing B_{n+1} long enough so that if $x \in B_n$, then $V(x) \subseteq B_n \cup B_{n+1}$. Then either the union of the even-indexed blocks is in $\mathcal{U}$ or the union of the odd-indexed blocks is in $\mathcal{U}$. Either way yields a set in $\mathcal{U}$ with no infinite path.

(11)$\Rightarrow$(12). If $V|X$ is independent, then $V|X$ is independent (mod $[\omega]^{<\omega}$).

We will now prove the ten reversals going from the bottom of the diagram to the top. After this, we prove the uniqueness assertions (i.e., the non-reversals for ultrafilters that are not of the type indicated).

(12)$\Rightarrow$(11) for Q-points. Suppose $V|X$ is independent (mod $[\omega]^{<\omega}$) for some set $X \in \mathcal{U}$. Consider the sequence $\langle B_n : n \in \omega \rangle$ of consecutive blocks of integers in X used in the proof of $9 \Rightarrow 10$ above. Again, either the union of the even-indexed blocks is in $\mathcal{U}$ or the union of the odd-indexed blocks is in $\mathcal{U}$. And because $\mathcal{U}$ is a Q-point, there exists a set $Y \in \mathcal{U}$ consisting of one point from each of the blocks. This yields the desired independent set in $\mathcal{U}$.

(11)$\Rightarrow$(10) for P-points. We show that for every V there exists a set $X \in \mathcal{U}$ such that $V|X$ is complete (mod $[\omega]^{<\omega}$) or $V|X$ is independent (mod $[\omega]^{<\omega}$). Actually, a much stronger result was announced in [BT78]. If $V \in \mathcal{U} \times \mathcal{U}$ then, because $\mathcal{U}$ is a P-point, we can find a set $X \in \mathcal{U}$ such that for every $x \in X$ we have $X \subseteq V(x)$ (mod $[\omega]^{<\omega}$). In this case, $V|X$ is complete (mod $[\omega]^{<\omega}$). Similarly, if $V^c \in \mathcal{U} \times \mathcal{U}$, then $V^c|X$ is complete (mod $[\omega]^{<\omega}$), and so $V|X$ is independent (mod $[\omega]^{<\omega}$).

(10)$\Rightarrow$(9). Assume $X \in \mathcal{U}$ and let Y be the infinite path in X. Then for every $y \in Y$ we have $V(y) \cap Y \neq \emptyset$. Thus X is not V-scattered.

(9)$\Rightarrow$(8). Assume that $B(X) \in \mathcal{U}$ but $X \notin \mathcal{U}$. We'll show that $B(X) - X$ is scattered. Assume that Y is a nonempty subset of $B(X) - X$. Chose $y \in Y$ such that $y \in B_\alpha(X)$ and α is minimal. Then $\alpha > 0$, and so $V(y) \cap Y = \emptyset$.

$(5) \Rightarrow (4)$. If $V \notin \mathcal{U} \times \mathcal{U}$, then $\{x : V(x) \notin \mathcal{U}\} \in \mathcal{U}$ (that is, a measure one set of agents have null visibility). If S is a $\mathcal{U}$-robust predictor, then any agent whose visibility is null is effectively blind, and it is easy to make all such agents guess incorrectly. So there is no $\mathcal{U}$-robust $\mathcal{U}$-predictor.

$(4 \Rightarrow 3)$. This was done in the proof of $(3) \Rightarrow (4)$.

We now conclude with proofs of the non-implications.

$(12) \not\Rightarrow (11)$ for non-Q-points. Suppose that f shows that $\mathcal{U}$ is not a Q-point, and let V be the graph where x and y are adjacent if they are distinct points with $f(x) = f(y)$. Then for all $X \in \mathcal{U}$, $V|X$ is not independent, but V itself is independent (mod $[\omega]^{<\omega}$).

$(11) \not\Rightarrow (10)$ for non-P-points. Suppose that f shows that $\mathcal{U}$ is not a P-point and let V be the graph where for $x < y$ we have an edge between them if $f(x) > f(y)$. Then for every $X \in \mathcal{U}$, $V|X$ is not independent (mod $[\omega]^{<\omega}$), but there is no infinite path in the graph V.

$(6) \not\Rightarrow (3)$ for non-P-points. Suppose that f shows that $\mathcal{U}$ is not a P-point and let V be the graph where for $x < y$ we have an edge between them if $f(x) = f(y)$. Then $V \notin \mathcal{U} \times \mathcal{U}$, but there is a $\mathcal{U}$-predictor obtained as follows. For each n, we let the agents in $f^{-1}(\{n\})$ use a robust finite-error strategy among themselves, ignoring all other agents. Then, for any coloring, the function f is finite-to-one on the set of errors and so this set is not in $\mathcal{U}$.

$(3) \not\Rightarrow (2)$ for non-P-points. Suppose that f shows that $\mathcal{U}$ is not a P-point and let V be the graph where for $x < y$ we have an edge between them if $f(x) < f(y)$. Then $V \in \mathcal{U} \times \mathcal{U}$, but for every $X \in \mathcal{U}$, some agent x fails to see infinitely many other agents in the set (namely those in $f^{-1}(\{f(x)\})$).

$(2) \not\Rightarrow (1)$ for non-Q-points. Suppose that f shows that $\mathcal{U}$ is not a Q-point and let V be the graph where for $x < y$ we have an edge between them if $f(x) = f(y)$. Then V itself is complete (mod $[\omega]^{<\omega}$), but $V|X$ is not complete for any $X \in \mathcal{U}$. $\square$

4.8 Blass's Evasion and Prediction Setting

Let $\mathbf{Z}^{\aleph_0}$ denote the direct product of denumerably many copies of the group $(\mathbf{Z}, +)$ and, for each $n \in \omega$, let e_n be the characteristic function of the set $\{n\}$. Ernst Specker [S50] proved that any homomorphism from $\mathbf{Z}^{\aleph_0}$ to $\mathbf{Z}$ maps e_n to 0 for all but finitely many n. He also showed that many of the subgroups of $\mathbf{Z}^{\aleph_0}$ with cardinality $2^{\aleph_0}$ share this property. This raises the question of whether or not such a subgroup must have cardinality $2^{\aleph_0}$, and this was answered by K. Eda [E83] who showed that the existence of such a subgroup of size $\aleph_1$ is independent of ZFC $+ 2^{\aleph_0} = \aleph_2$.

The connection between these algebraic investigations of Specker and Eda and the kind of predictors that we are interested in here dates from the 1994 work of Andreas Blass [Bla94] and some initial results he obtained that relate the minimal size of a subgroup of $\mathbf{Z}^{\aleph_0}$ satisfying Specker's theorem to several of the standard cardinal characteristics of the continuum such as those that occurred in Sect. 4.5.

The fourth section of Blass's paper was entitled "Predicting and Evading," and this is the area that we wish to preview here.

Throughout this chapter, "one-way visibility on ω" has meant that agents never see the hats worn by smaller-numbered agents—that is, "visibility to the right (or up)." We never speak of "visibility to the left (or down)" on ω because we know that with finite visibility, there is no minimal predictor.

Blass, however, was led to precisely the situation in which the set of agents (and the set of colors) is ω and agent n sees the hats worn by agents 0 through $n-1$. The assertion that there is no minimal predictor here implies that for every predictor P, there is a "counterexample" $h \in {}^{\omega}\omega$ showing that P is not a minimal predictor, and thus that infinitely many agents in X guess incorrectly for h for every $X \in [\omega]^{\omega}$. But how large a "pool" of potential counterexamples from ${}^{\omega}\omega$ do we need to ensure that for each predictor we have at least one such counterexample? This question leads to the following (extracted from [Bla94]).

Definition 4.8.1. A family $\mathcal{E} \subseteq {}^{\omega}\omega$ is an *evading family* if for every predictor P for the hat problem with ω colors, ω agents, and full visibility to the left, and for every $X \in [\omega]^{\omega}$, there exists a coloring $h \in \mathcal{E}$ for which infinitely many agents in X guess incorrectly. The *evasion number* $\mathfrak{e}$ is the smallest possibility cardinality of an evading family.

Blass obtained a number of results relating $\mathfrak{e}$ to some of the known characteristics of the continuum, as well as to some of those arising from his original algebraic setting, and a nice summary of these can be found in Sect. 10 of [Bla10]. This work led to a series of papers by a number of other authors, and while all of these results are interesting and important, they carry us a bit far afield of our present considerations. Nevertheless, for the sake of the reader who wants to pursue this line, we collect together here, in chronological order, a number of these contributions by title and author.

- 1994: Cardinal characteristics and the product of countably many infinite cyclic groups, Blass [Bla94].
- 1995: Evasion and prediction – the Specker phenomenon and Gross spaces, Brendle [Bre95].
- 1996: Evasion and prediction II, Brendle and Shelah [BS96].
- 1998: The Baire category theorem and the evasion number, Kada [Kad98].
- 2000: Cardinal invariants associated with predictors, Kamo [Kam00].
- 2001: Cardinal invariants associated with predictors II, Kamo [Kam01].
- 2003: Evasion and prediction III: Constant prediction and dominating reals, Brendle [Bre03].
- 2003: Evasion and prediction IV: Strong forms of constant prediction, Brendle and Shelah [BS03].

4.9   Open Questions

The problem of characterizing the amount of one-way visibility needed for an optimal predictor on ω and for a minimal predictor on ω was completely solved for transitive graphs in Sect. 4.2. For nontransitive graphs on ω, the characterization problem for optimal predictors on ω was solved in Sect. 4.3. But the corresponding result for minimal predictors on ω is notably absent, and it seems that the number of colors may play a prominent role here.

For the case where we seek enough one-way visibility to get a minimal predictor on ω for an arbitrarily large set of colors, we begin with the observation that an infinite complete subgraph certainly suffices. That is, these agents then have a finite-error predictor among themselves and this ensures infinitely many correct guesses.

Question 4.9.1. Suppose V is a graph on ω providing enough visibility for the existence of a predictor that ensures at least one correct guess for an arbitrarily large set of colors. Must V contain an infinite complete subgraph?

For questions with regard to the parity relation EO, we begin with a piece of terminology and an observation.

Definition 4.9.1. Say that $\langle f_\alpha : \alpha < \omega_1 \rangle \subseteq {}^\omega v$ is a *strongly agreeable family* if $(\forall g \in {}^\omega v)(\exists \alpha \in \omega_1)(\forall \beta > \alpha)$ the functions f_α and g agree infinitely.

Theorem 4.9.2. *The following are equivalent:*

1. *There exists a strongly agreeable family in ${}^\omega \omega_1$.*
2. *There is a predictor that ensures at least one correct guess for the parity relation EO with ω_1 colors.*

Proof. $(1) \Rightarrow (2)$: Fix a strongly agreeable family $\langle f_\alpha : \alpha < \omega_1 \rangle$ in ${}^\omega \omega_1$. The predictor is as follows. For a hat assignment $h = (h^0, h^1)$, the agents in A^i choose α^i such that $(\forall \beta \geq \alpha^i)$ h^{1-i} and f_β agree infinitely, and guess according to f_{α^i}.

We must have $\alpha^0 \leq \alpha^1$ or $\alpha^1 \leq \alpha^0$. If $\alpha^{1-i} \leq \alpha^i$, then since h^i agrees infinitely with f_β for $\beta \geq \alpha^{1-i}$, h^i agrees infinitely with f_{α^i}, so infinitely many agent in P^i guess correctly.

$(2) \Rightarrow (1)$: Suppose there is no strongly agreeable family in ${}^\omega \omega_1$. Fix any predictor $P = (P^0, P^1)$. Let c_α be as in the proof of Theorem 4.4.2. Let $f_\alpha = S^1(c_\alpha)$. Since $\langle f_\alpha : \alpha < \omega_1 \rangle$ is not a strongly agreeable family, we can choose $g \in {}^\omega \omega_1$ such that $(\forall \alpha \in \omega_1)(\exists \beta > \alpha)$ g only finitely agrees with f_β. In particular, there exists $\beta > \sup P^0(g)$ such that g only finitely agrees with f_β. Then, under the hat assignment (c_β, g), no agent in A^0 guesses correctly since $\beta > \sup P^0(g)$, and only finitely many agents in A^1 guess correctly since g only finitely agrees with $f_\beta = P^1(c_\beta)$. This establishes that there is no predictor ensuring infinitely many correct guesses, so there is no minimal predictor. $\square$

In an early draft of this manuscript we asked if the existence of a minimal predictor for the parity relation EO with $\aleph_1$ colors implied CH. An anonymous referee pointed out that if one simply adds $\aleph_2$ Cohen reals to a model of ZFC + CH, then there is a strongly agreeable family $\langle\, f_\alpha : \alpha < \omega_1 \,\rangle$ in ${}^\omega\omega_1$. In fact (as also pointed out by the referee and included with their permission), one can also do the following. Start with any model of ZFC in which the continuum hypothesis is false and force with the direct sum $\sum_{\alpha<\omega_1} \mathbf{P}_\alpha$ (finite support product) where each $\mathbf{P}_\alpha$ is the countable partial order of finite partial functions from ω to α. Let $f_\alpha : \omega \to \alpha$ be the generic real added by $\mathbf{P}_\alpha$. Let $V_\alpha = V[f_\beta : \beta < \alpha]$ be the generic extension up to stage $\alpha \leq \omega_1$. It follows from standard arguments about ccc forcing that for any $g : \omega \to \omega_1$ in V_{ω_1} there will exist $\alpha < \omega_1$ with $g \in V_\alpha$ and α containing the range of g. For any $\beta > \alpha$, it follows from the product lemma that f_β will be generic over V_β which contains V_α. Hence $f_\beta \cap g$ will be infinite.

How much visibility is needed for a minimal predictor to exist? Of course, this depends on the model, so we could ask two questions: how much visibility is needed for a minimal predictor to exist under CH? How much is needed in the model where we've added $\aleph_2$ random reals?

From the section on $\mathcal{U}$-predictors, the following remains unresolved.

Question 4.9.2. Is it true that for every non-P-point $\mathcal{U}$, there exists a graph V on ω such that for all $X \in \mathcal{U}$, $V|X$ has an infinite path, but there exists no $\mathcal{U}$-predictor for V?

Chapter 5
Dual Hat Problems, Ideals, and the Uncountable

5.1 Background

There are a number of notions of "dual" in mathematics, many of which fit into the framework provided by category theory. The notion of dual that we are interested in here is perhaps most evident in the following formulation:

(a) A function $f : B \to C$ is injective iff for every set A and every pair of functions $g, h : A \to B$, if $f \circ g = f \circ h$, then $g = h$.
(b) A function $f : A \to B$ is surjective iff for every set C and every pair of functions $g, h : B \to C$, if $g \circ f = h \circ f$, then $g = h$.

There is also a natural identification between subsets of A and injective functions mapping an ordinal to A (although different injective functions correspond to the same subset), and between partitions of A and surjective functions mapping A to an ordinal (with the same caveat). With this identification in mind, we are interested in the notion of dual that links injective functions and surjective functions or, equivalently, links subsets and partitions.

Known results suggest that theorems pertaining to injections or subsets tend to be "weaker" than the corresponding theorems in the context of surjections or partitions. We will give two examples to illustrate this, one in the context of injections and surjections and one in the context of subsets and partitions. We do this to set the stage for Sect. 5.2 where we consider duals in the context of hat problems.

For our first example, recall that the Schroeder-Bernstein theorem asserts that for any two sets X and Y, if there are injections from X to Y and from Y to X, then $|X| = |Y|$. Moreover this is a theorem of ZF. The so-called *dual Schroeder-Bernstein theorem* asserts that for any two sets X and Y, if there are surjections from X to Y and from Y to X, then $|X| = |Y|$. While this is a theorem of ZFC, it is not a theorem of ZF + DC; see [BM90].

For our second example, we start with the finite version of Ramsey's theorem in the form asserting that for any two positive integers k and r there exists a positive integer $n = R(k, r)$ such that for every r-coloring c of the subsets of n, there exists

C.S. Hardin and A.D. Taylor, *The Mathematics of Coordinated Inference*,
Developments in Mathematics 33, DOI 10.1007/978-3-319-01333-6_5,
© Springer International Publishing Switzerland 2013

a subset H of n of size k that is homogeneous for c in the sense that the color of any subset of H depends only on its size. The *dual Ramsey theorem*, originally obtained by Ronald Graham and Bruce Rothschild [GR71], replaces "subset (of size k)" with "partition (consisting of k sets)" and "subset of the subset H of n" with "coarser partition than the partition H of n." The dual Ramsey theorem is quite powerful, yielding both Ramsey's theorem itself and the celebrated theorem of van der Waerden on arithmetic progressions. There are also infinite versions of duals; e.g., see [CS84] for a discussion of what they call the "dual Ellentuck theorem."

After dealing with dual hat problems in Sect. 5.2, we consider extending some of the results in Chap. 4 to the uncountable, and we do this in Sect. 5.3 in the context of ideals on a cardinal. This leads to the role played by non-regular ultrafilters in Sect. 5.4 and a hat problem whose solution is equivalent to the GCH in Sect. 5.5.

5.2 Dual Hat Problems

In Chap. 1, we described a "reasonably general framework" for hat problems as being made up of a set A of agents, a set K of colors, a set C of colorings (each mapping A to K), and a collection $\{\equiv_a: a \in A\}$ of equivalence relations on C with the interpretation being that $f \equiv_a g$ indicates that agent a cannot distinguish between the colorings f and g. In all of our applications so far, the equivalence relations have corresponded, at least metaphorically, to myopia (near-sightedness). That is, we've typically had $f \equiv_a g$ because $\{b \in A : f(b) \neq g(b)\}$ is a set of agents not seen (or at least not clearly seen) by agent a.

In the case of a visibility graph, we can explicitly define the corresponding equivalence relation in terms of injections as follows: There exists, for each $a \in A$, an ordinal α_a and an injection $h_a : \alpha_a \to A$ so that for $f, g \in C$, we have $f \equiv_a g$ iff $f \circ h_a = g \circ h_a$. Intuitively, the dual of the situation $\alpha_a \to A \to K$ with the first map injective would be $A \to K \to \alpha_a$ with the second map surjective. Thus, the mathematical notion of a dual would suggest looking at equivalence relations $\equiv_a$ on C derived from surjections $h_a : K \to \alpha_a$ by $f \equiv_a g$ iff $h_a \circ f = h_a \circ g$.

Interestingly, this notion of indistinguishable colorings makes sense in the metaphorical context. The issue is no longer having other players beyond the range of agent a's vision, but that agent a cannot distinguish between (or among) certain colors. For example, he may be red-green colorblind!

It turns out that the μ-predictor has something to say in the dual hat problem as well. To see this, suppose that A and K are arbitrary, $C = {}^A K$ and we have a collection $\{h_a : a \in A\}$ where h_a is a surjection from K to some ordinal α_a. Define $\equiv_a$ on C by $f \equiv_a g$ iff $h_a \circ f = h_a \circ g$.

Here, we will re-interpret a guessing strategy G_a for agent a to be a map from C to C (as opposed to a map from C to K), and we will similarly re-interpret the μ-predictor such that if $\prec$ is a well ordering of C, then the guessing strategy $\mu_a(f)$ for agent a is the $\prec$-least g such that $f \equiv_a g$. Notice that we have agent a guessing the whole coloring and not just his own hat color. How successful is this predictor?

The μ-predictor's success in the earlier contexts was, at least in part, tied to transitivity of the visibility graph. And one way to phrase transitivity, at least in the case where the set A of agents is an ordinal, is to say that $\alpha < \beta$ implies that agent α's knowledge of a coloring is a superset of agent β's knowledge of the coloring. This guarantees that if we have an ω-sequence $\alpha_0 < \alpha_1 < \cdots$ of agents using the μ-strategy, then $\langle f \rangle_{\alpha_0} \succeq \langle f \rangle_{\alpha_1} \succeq \cdots$. Because $\prec$ is a well-ordering it follows that for some n we have $\langle f \rangle_{\alpha_n} = \langle f \rangle_{\alpha_{n+1}} = \cdots$.

Now, still in the earlier context, if agent α_n sees agent α_{n+1} and agent α_{n+1} is trying to guess his own hat color, then α_{n+1} is trying to guess something that agent α_n already knows. One way to say that is to define agent β's guess to be *acceptable* if there exists some $\gamma < \beta$ such that agent β's guess is consistent with what agent γ knows. With this definition, our discussion yields a proof of the following.

Theorem 5.2.1. *Suppose that the set A of agents is an ordinal α, the set K of colorings is arbitrary, $C = {}^A K$, and we have a collection $\{ h_\beta : \beta \in A \}$ where h_β is a surjection from K to some ordinal and such that if $\gamma < \beta$ and $h_\gamma(k_1) = h_\gamma(k_2)$ then $h_\beta(k_1) = h_\beta(k_2)$. Consider the hat problem in which $f \equiv_\beta g$ iff $h_\beta \circ f = h_\beta \circ g$ and for which $G_\beta(f)$ is acceptable if there exists some $\gamma < \beta$ such that $f \equiv_\gamma G_\beta(f)$. Then for any coloring, the μ-predictor guarantees unacceptable guesses from only finitely many agents.*

Dual hat problems represent an area largely unexplored. Nevertheless, we leave them now and turn to a consideration of ideals on uncountable cardinals and on ω.

5.3 Hat Problems and Ideals

In Sect. 4.2 we considered the problem of characterizing those transitive visibility graphs on ω that yield a finite-error predictor and those that yield a minimal predictor, and Theorems 4.2.1 and 4.2.2 solved these problems. We begin this section by generalizing those results to the context of ideals on uncountable cardinals. As in Sect. 4.2 we only consider one-way visibility in this section.

If I is an ideal on the infinite cardinal κ, then the notation $I^+ \rightarrow (I^+, \omega)^2$ denotes the assertion that for every set $X \in I^+$ and every function $f : [X]^2 \rightarrow 2$, there exists a set $Y \subseteq X$ such that either $Y \in I^+$ and $f([Y]^2) = 0$ or $|Y| = \omega$ and $f([Y]^2) = 1$. Ramsey's theorem asserts that $I^+ \rightarrow (I^+, \omega)^2$ when $I = [\omega]^{<\omega}$, the Dushnik-Miller-Erdős theorem [EHMR84] asserts that $I^+ \rightarrow (I^+, \omega)^2$ when $I = [\kappa]^{<\kappa}$ for any (infinite) cardinal κ, and it is well known that if κ is regular, then $NS_\kappa^+ \rightarrow (NS_\kappa^+, \omega)^2$, where NS_κ is the ideal of nonstationary subsets of κ.

Theorems 4.2.1 and 4.2.2 showed that if $I = [\omega]^{<\omega}$ and V is a transitive graph on ω, then there exists a positive I-measure predictor iff V contains an infinite complete subgraph, and there exists an I-measure one predictor iff V contains no infinite independent subgraph. The following generalizes this to transitive graphs and arbitrary ideals on an uncountable cardinal.

Theorem 5.3.1. *Suppose that I is an ideal on κ and V is an undirected transitive graph on κ. Consider the hat problem with one-way visibility given by V. Then (1) and (2) are equivalent, (3) implies (4), and, if $I^+ \to (I^+, \omega)^2$, then (4) implies (3) and so they too are equivalent.*

1. *There exists a positive I-measure predictor for two colors.*
2. *Every set of I-measure one contains an infinite complete subgraph.*
3. *There exists an I-measure one predictor for two colors.*
4. *There is no independent set of positive I-measure.*

Proof. (1)$\Rightarrow$(2). Assume (2) fails and let $X \subseteq \kappa$ be a set of positive I-measure with no infinite complete subgraph. Because V is transitive, this means that X contains no infinite path. Thus, if P is a predictor, then we can produce a 2-coloring $f : \kappa \to 2$ for which every agent in X guesses incorrectly exactly as in the proof of Theorem 4.2.2.

(2)$\Rightarrow$(1). We use the μ-predictor. Suppose for contradiction that $X \in I^*$ and everyone in X guesses incorrectly for some coloring. By (2) we can choose an infinite complete subgraph in X. But this now yields, as usual, an infinite descending chain in the well-ordering of the colorings.

(3)$\Rightarrow$(4). If (4) fails and $X \subseteq \kappa$ is an independent set of positive I-measure, then for every predictor we know, by Corollary 1.3.3, that there is a 2-coloring $f : \kappa \to 2$ for which every agent in X guesses incorrectly.

(4)$\Rightarrow$(3) assuming $I^+ \to (I^+, \omega)^2$. We use the μ-predictor. Suppose for contradiction that $X \in I^+$ and everyone in X guesses incorrectly for some coloring. Define $h : [X]^2 \to 2$ by $h(\gamma, \beta) = 0$ iff neither γ nor β can see the other. Because $I^+ \to (I^+, \omega)^2$, we get a subset Y of X such that either Y is an independent set of positive I-measure or an infinite complete subgraph. The former is ruled out by (4) and the latter yields, as usual, an infinite descending chain in the well-ordering of the colorings. $\square$

Corollary 5.3.2. *Suppose that V is an undirected transitive graph on κ, and consider the hat problem with two colors and one-way visibility given by V. Then there exists a $<\kappa$-error predictor iff there is no independent set of size κ. If κ is regular, then there exists a predictor ensuring the set or errors is nonstationary iff there is no stationary independent set.*

We now turn our attention to ideals on ω. Recall that an ideal I on ω is an NIS ideal if there is an I-measure one predictor whenever visibility is given by a graph V on ω having no independent set in I^+. In Chap. 4 we showed that:

• The ideal $I = [\omega]^{<\omega}$ is an NIS ideal (Theorem 4.3.1).
• If I is a maximal ideal (that is, if I^* is an ultrafilter), then I is an NIS ideal iff I^* is a Ramsey ultrafilter (Theorem 4.6.5).

In the remainder of this section, we derive some relationships among NIS ideals and a few of the well-known classes of ideals on ω dating back to [G71] and extended to the uncountable in [BTW82]. For the remainder of this section, we will assume that there are exactly two colors. Our starting point is the following.

Definition 5.3.3. An ideal on ω is said to be:

(a) a *local P-point* if every $f \in {}^{\omega}\omega$ is either constant on a set in I^+ or finite-to-one on a set in I^+;
(b) a *local Q-point* if every finite-to-one $f \in {}^{\omega}\omega$ is one-to-one on a set in I^+;
(c) *locally selective* if every $f \in {}^{\omega}\omega$ is either constant on a set in I^+ or one-to-one on a set in I^+; and
(d) *selective* if every $f \in {}^{\omega}\omega$ either has finite range on a set in I^* or is one-to-one on a set in I^+.

There are natural non-examples of local P-point ideals on ω and local Q-point ideals on ω. Both examples begin with a partition $\mathcal{P} = \langle A_n : n \in \omega \rangle$ of ω. For a non-local P-point, we require all the sets in $\mathcal{P}$ to be infinite, and define I as the collection of sets intersecting all but finitely many of the sets A_n in a finite set. For a non-local Q-point, we require all the sets in $\mathcal{P}$ to be finite, but with $\sup\{|A_n| : n \in \omega\} = \omega$, and we define J as the collection of sets X for which $\sup\{|X \cap A_n| : n \in \omega\} < \omega$. In fact, it is not hard to see that an ideal is a local P-point iff it contains no ideal isomorphic to the ideal I just constructed and it is a local Q-point iff it contains no ideal isomorphic to the ideal J just constructed.

If I is an ideal on ω and $A \in I^+$, then $I|A$ denotes the ideal of subsets of ω whose intersection with A is in I. That is, $I|A = \{X \subseteq \omega : X \cap A \in I\}$. The ideal $I|A$ is often called "the restriction of I to A," but it can also be described as the ideal generated by I and the one set A^c. If an ideal has one of the first three properties described above, there is no guarantee that a restriction of the ideal also has the property. For this reason, we introduce the following.

Definition 5.3.4. An ideal I is a *weak P-point* if $I|A$ is a local P-point for every $A \in I^+$. Similarly, I is a *weak Q-point* if $I|A$ is a local Q-point for every $A \in I^+$, and I is a *weakly selective ideal* if $I|A$ is a locally selective ideal for every $A \in I^+$.

It is now easy to see that an ideal is weakly selective iff it is both a weak P-point and a weak Q-point (although the analogous statement with the local version is false). Moreover, if I is a selective ideal, then so is $I|A$ for every $A \in I^+$. Selective ideals, of course, are easily seen to be weakly selective.

There is a characterization of weakly selective ideals that is relevant to the issue of obtaining predictors, and is implicit in [BT78]. It asserts that an ideal I on ω is weakly selective iff $I^+ \to (I^+, \omega)^2$. We sketch the proof. First, if I is weakly selective and V is a graph with vertex set $A \in I^+$, then we consider two cases. If for every $B \in \mathcal{P}(A) \cap I^+$ there is some $b \in B$ such that $V(b) \cap B \in I^+$, then it is easy to build an infinite complete subgraph. If this fails for some $B \in I^+$, then we can use the weak-P-pointness of I to get a set of positive I-measure upon which visibility is finite. But now it is easy to get a sequence of consecutive blocks of numbers so that every agent in one block sees only agents in his own block or the next one. To get the desired independent set of positive I-measure, we now use the weak-Q-pointness of I together with the observation that either the union of the odd-numbered blocks is in I^+ or the union of the even-numbered blocks is. For the converse, the partition relation can be used to show that I is a weak P-point and a weak Q-point.

There is also a weakened notion of NIS ideal that is relevant to our present discussion.

Definition 5.3.5. An ideal I on ω is a *local NIS ideal* if there is a minimal predictor (ensuring at least one correct guess) whenever visibility is given by a graph V on ω having no independent set in I^+. It is a *weak NIS ideal* if $I|A$ is a local NIS ideal for every $A \in I^+$.

Notice that if I is a local NIS ideal, then so is every subideal of I (a property that also holds for local P-points, local Q-points, and locally selective ideals). Thus, for example, if I^* can be extended to a Ramsey ultrafilter, then I is a local NIS ideal.

For selective ideals, there is no need to define a local version, because if I is selective, then so is $I|A$ for every $A \in I^+$. A similar thing is true for NIS ideals.

Proposition 5.3.6. *If I is an NIS ideal on ω, then so is $I|A$ for every $A \in I^+$.*

Proof. Suppose V has no independent set in $(I|A)^+$ where $A \in I^+$. Let V' be the graph obtained by adding edges between every pair of vertices in A^c. If X is an independent set in V', then X is independent in V and $|X \cap A^c| \leq 1$. Thus, $X \in I|A$ and so $X \cap A \in I$. Because I is an NIS ideal and every independent set in V' is in I, we know that there exists an I^*-predictor for V'. Notice that for every agent $a \in A$, the set of agents visible to agent a is unchanged in passing from V to V'. Hence, we can arrive at the desired predictor P for V simply by having the agents in A use the predictor P' (and letting the agents in A^c guess arbitrarily). Only a set of agents of I-measure zero guess incorrectly with P' so the set of agents guessing incorrectly with P is in $I|A$. $\qquad\square$

More to the point, we have the following.

Theorem 5.3.7. *Every weak NIS ideal I on ω is a weakly selective ideal.*

Proof. It suffices to show that every weak NIS ideal on ω is both a local P-point and a local Q-point, as this will then imply that every weak NIS ideal is a weak P-point and a weak Q-point, and thus weakly selective as desired.

To see that I is a local P-point, suppose that $f \in {}^\omega\omega$. Let V be the visibility graph on ω where for $n < m$, we have that n is adjacent to m iff $f(n) > f(m)$. Notice first that Proposition 1.3.2 guarantees there is no minimal predictor for V because agents in $f^{-1}(\{0\})$ are blind, agents in $f^{-1}(\{1\})$ see only agents in $f^{-1}(\{0\})$, etc. Since I is a weak NIS ideal, it follows that there must be an independent set $X \in I^+$. If $f|X$ is finite-to-one, then we are done. If not, choose i such that $f^{-1}(\{i\})$ is infinite. Then, because X is independent, it follows that $f^{-1}(\{j\}) = \emptyset$ for every $j > i$. Thus $X \subseteq f^{-1}(\{0\}) \cup \cdots \cup f^{-1}(\{i-1\})$, and so $f^{-1}(\{p\}) \in I^+$ for some $p \leq i$.

To see that I is a local Q-point, suppose that $f \in {}^{\omega}\omega$ is finite-to-one. Let V be the visibility graph on ω where for $n \neq m$, we have that n is adjacent to m iff $f(n) = f(m)$. Because visibility is finite with V and we only have two colors, Proposition 1.3.7 guarantees that there is no minimal predictor for V. Hence, because I is an NIS ideal, there is an independent set $X \in I^{+}$. But clearly f is constant on any independent set. $\square$

There is a weak converse to the above result. Adrian Mathias [Mat77] has shown that if CH holds and I is a selective ideal on ω, then I^{*} can be extended to a Ramsey ultrafilter on ω. And we know that Ramsey ultrafilters are NIS ideals, and thus local NIS ideals. As subideals of local NIS ideals are again local NIS ideals, it follows that I is a local NIS ideal. The same argument can be applied to $I|A$ for any $A \in I^{+}$. Thus, CH implies that every selective ideal on ω is a weak NIS ideal.

We conclude with one more class of ideals.

Definition 5.3.8.   An ideal I on ω is a *TR ideal* if for every graph $V = (\omega, E)$ with no independent set in I^{+} there exists a set $X \in I^{*}$ and a set $E' \subseteq E$ such that (X, E') has no independent set in I^{+} and E' is transitive.

We showed in Chap. 4 that $I = [\omega]^{<\omega}$ is a TR ideal (and you can always take $X = \omega$), and if I is dual to a Ramsey ultrafilter then I is a TR ideal (and you can always take $E' = E$). If I is a TR ideal on ω and $I^{+} \rightarrow (I^{+}, \omega)^2$ then, because of the transitivity of E', we can use the μ-predictor with assurances that the set of errors will be of I-measure zero as long as there is no independent set of positive-I-measure. This, together with our earlier comments, yields the following.

Proposition 5.3.9.   *Every weakly selective TR ideal on ω is an NIS ideal.*

5.4   The Role of Non-regular Ultrafilters

In the absence of transitivity, Theorem 4.3.1 nevertheless characterizes those visibility graphs on ω that are adequate to yield finite-error predictors: they are precisely the ones containing no infinite independent subgraph. The proof of this result proceeded by inductively throwing away edges until one arrives at a transitive graph, and doing this in a way that guarantees that no infinite independent subgraph is created in the process. It turns out, however, that this approach will not generalize to all larger cardinals, as it was shown in [H10] that there is a graph on $2^{\aleph_0}$ with no infinite independent subgraph, but all of whose transitive subgraphs contain an infinite independent subgraph.

Theorem 4.3.1 also does not generalize in the obvious way to countably infinite directed graphs. For example, the directed graph on ω where there is an edge from each number to each smaller number has no infinite independent subgraph but no predictor can ensure even one correct guess.

However, if we begin with an undirected graph V on a countable ordinal α and consider the hat problem with one-way visibility given by V, then Theorem 4.3.1 does extend in the obvious way. That is, letting $f : \omega \to \alpha$ be a bijection, we consider the corresponding graph V' on ω that is isomorphic to V via f. Visibility on ω is then given by n sees m iff $n < m$, $f(n) < f(m)$, and there is an edge in V between $f(n)$ and $f(m)$. It now follows that V' has no infinite independent set X because we could then partition the pairs from X according to whether the orderings from ω and α agree on the pair or not. An infinite set homogeneous for disagreement would yield an infinite descending chain in α, and an infinite set homogeneous for agreement would yield an infinite independent set in the graph V. Hence, we can now invoke Theorem 4.3.1 to get a finite-error predictor for V' and this immediately yields a finite-error predictor on α with visibility given by V.

The observation in the preceding paragraph suggests an approach for obtaining a reasonably successful predictor on ω_1 from a visibility graph with no large independent sets. That is, if V is a graph on ω_1 with no infinite independent subgraph, then for every $\alpha < \omega_1$ we can consider the restriction V_α of V to α and get a finite-error predictor P_α that equips each agent $\beta < \alpha$ with a strategy. Now to get a predictor for ω_1, we want to amalgamate these predictors in some way. Perhaps the most natural amalgamation is to use an ultrafilter $\mathcal{U}$ on ω_1 and have agent β guess red iff $\{ \alpha > \beta : P_\alpha \text{ has agent } \beta \text{ guess red} \} \in \mathcal{U}$.

It seems, however, that to get any traction from this approach requires a rather special—but well known—kind of ultrafilter on ω_1.

Definition 5.4.1. An ultrafilter on ω_1 containing no countable sets is *non-regular* if for every collection of uncountably many sets from the ultrafilter, some infinite subcollection has nonempty intersection.

While one cannot prove in ZFC that non-regular ultrafilters on ω_1 exist, it is known that relative to the existence of a sufficiently large cardinal, it is consistent that they exist. Their use in our context is the following.

Theorem 5.4.2. *Assume there exists a non-regular ultrafilter on ω_1. Suppose that V is an undirected graph on ω_1 with no infinite independent subgraph, and consider the hat problem with finitely many colors and one-way visibility given by V. Then there exists a predictor ensuring at most countably many incorrect guesses.*

Proof. If $\mathcal{U}$ is the postulated non-regular ultrafilter, then we use the amalgamation described above to arrive at our predictor. Suppose, for contradiction, that f is a coloring and uncountably many agents guess incorrectly for f. Each such agent β guessed based on a set $X_\beta \in \mathcal{U}$. Non-regularity now yields $\beta_0 < \beta_1 < \cdots < \alpha$ and $\alpha \in X_{\beta_i}$ for each $i < \omega$. Thus P_α has agent β_i guessing the same as the predictor on ω_1 so P_α has infinitely many incorrect guesses, a contradiction. $\square$

5.5 A Hat Problem Equivalent to the GCH

We consider here a generalization of the standard hat problem in which each agent tries to guess the whole coloring. So here, as in Sect. 5.2, a guessing strategy G_a for an agent will be a function from colorings to colorings, but the guess will be deemed correct only if $G_a(f) = f$. That is, we are interested in how many agents can guess the whole coloring based on what they see. Of course, if every agent could see every other agent, this would be the same as asking them to guess their own hat color. But we are working with undirected graphs, so it is only one-way visibility with no agent seeing the hats worn by smaller-numbered agents.

Can we find a predictor ensuring that infinitely many agents guess the whole coloring correctly? Or maybe a cofinite set of agents? And does the number of colors make any difference? The following theorem answers all of these questions for the case where the set of agents is denumerable.

Theorem 5.5.1. *Consider the hat problem on ω in which an agent a's guess is a coloring and this guess is considered correct only if $G_a(f) = f$. Then:*

1. *With a countable set of colors there is a predictor ensuring that infinitely many agents guess correctly, as long as there is no infinite independent set in the visibility graph.*
2. *With an uncountable set of colors there is no predictor ensuring that one agent will guess correctly, even with full visibility to the right.*
3. *With two colors and full visibility to the right, there is no predictor that will ensure a cofinite set of correct guesses. In fact, any predictor ensuring at least one correct guess for every coloring will also have infinitely many incorrect guesses for every coloring.*

Proof. Assume $A = \omega$, $K = \omega$, and $C = {}^{\omega}\omega$. Let $\langle s_n : n \in \omega \rangle$ enumerate the set of all functions mapping some finite subset of ω to ω such that each function occurs infinitely often. For each $n \in \omega$, our guessing strategy $G_n : C \to C$ will be the following modification of the μ-predictor: $G_n(f)(k) = s_n(k)$ if k is in the domain of s_n, and $G_n(f)(k) = \langle f \rangle_n(k)$ otherwise.

Given f, let p be the smallest natural number such that $\langle f \rangle_p = \langle f \rangle_{p+1} = \cdots$. Such a p must exist or we'd have an infinite descending chain in our well-ordering of C. Let $s = f|(p + 1)$. For any k we can choose $n > k$ such that $s_n = s$. We claim that $G_n(f) = f$. First, if $t \leq p$ then t is in the domain of s and thus the domain of s_n, so $G_n(f)(t) = s_n(t) = f(t)$. If $t > p$, then $G_n(f)(t) = \langle f \rangle_n(t) = \langle f \rangle_p(t) = f(t)$. This proves (1).

Now assume that the set of colors is uncountable, and P is some predictor. Let $f(k) = 0$ for all k and choose a color γ such that for every $n \in \omega$, $G_n(f)(0) \neq \gamma$. Let g be the coloring that agrees with f everywhere except that $g(0) = \gamma$. Then no one guesses g correctly. This proves (2).

Finally, suppose there are two colors, P is a predictor and $f \in {}^{\omega}2$ is such that $G_n(f) = f$ for all $n \geq m$, and m is the smallest natural number for which this is true. Let $g(k) = f(k)$ for $k \neq m$, and let $g(m) \neq f(m)$. Then $G_n(g) \neq g$ for any

$n \geq m$. So now we can start at $m - 1$ and work our way down to 0, changing hats so that each of these agents guesses incorrectly.   □

When we consider the same question in the context of the uncountable, we find that the solution rests on the GCH.

It is well known that the GCH is equivalent to the assertion that $\omega^{<\kappa} = \kappa$ for every infinite cardinal κ. To see this, note that the GCH immediately implies that for any $\lambda < \kappa$, $\omega^\lambda = \lambda^+ \leq \kappa$, so $\omega^{<\kappa} \leq \kappa$. It is elementary that $\omega^{<\kappa} \geq \kappa$, so $\omega^{<\kappa} = \kappa$. On the other hand, if the GCH fails, we can fix some λ with $2^\lambda > \lambda^+$ and let $\kappa = \lambda^+$. Then $\omega^{<\kappa} = 2^\lambda > \kappa$.

Theorem 5.5.2. *For any infinite cardinal κ, the following are equivalent.*

1. *There exists a predictor for the hat problem with denumerably many colors, κ agents, and full visibility to the right that guarantees κ-many agents guess the whole coloring correctly.*
2. *There exists a predictor for the hat problem with two colors, κ agents, and full visibility to the right that guarantees κ-many agents guess the whole coloring correctly.*
3. $\omega^{<\kappa} = \kappa$

Proof. $(1) \Rightarrow (2)$ is trivial.

For $(3) \Rightarrow (1)$, let s_α be a κ-sequence in which each element of $^{<\kappa}\omega$ appears κ-many times. We let player α guess that the coloring is the function

$$g(\beta) = \begin{cases} s_\alpha(\beta) & \text{if } \beta \in \text{dom}(s_\alpha), \\ \langle f \rangle_\alpha(\beta) & \text{otherwise.} \end{cases}$$

It is straightforward to verify that κ-many players guess the whole coloring correctly under this predictor.

For $(2) \Rightarrow (3)$, note that $2^{<\kappa} = \omega^{<\kappa} \geq \kappa$ and suppose $2^{<\kappa} > \kappa$. Let P be any predictor for the hat problem of (2). Since $2^{<\kappa} = \sup_{\lambda < \kappa} 2^\lambda > \kappa$, there is some infinite cardinal $\lambda < \kappa$ with $2^\lambda > \kappa$. We define the coloring f as follows. Let f be arbitrary on $[\lambda, \kappa)$. Then the guesses of all agents $\alpha \geq \lambda$ are determined, and since $2^\lambda > \kappa$, we can define $f \upharpoonright \lambda$ in a way that makes every $\alpha \geq \lambda$ guess incorrectly.   □

Corollary 5.5.3. *The GCH holds iff for every infinite cardinal κ there exists a predictor ensuring that κ agents correctly guess the whole coloring in the hat problem with either two colors or ω colors and full visibility of larger agents.*

As a final comment with regard to guessing the whole coloring, consider the hat problem with two colors on ω_1. The combinatorial principle $\Diamond_{\omega_1}$ is the assertion that there is a sequence $\langle A_\alpha : \alpha \in \omega_1 \rangle$ such that $A_\alpha \subseteq \alpha$ and for every $X \subseteq \omega_1$, we have that $\{ \alpha : X \cap \alpha = A_\alpha \}$ is a stationary set. If $\Diamond_{\omega_1}$ holds and each agent $\alpha \in \omega_1$ sees a closed unbound set of larger agents, then the Gabay-O'Connor predictor can be modified by having agent α guess that the coloring of hats worn by smaller players is given by A_α. With this, we are ensured of having a stationary set of agents guess the whole coloring correctly. This observation essentially originated with Gabay and O'Connor.

5.6 Open Questions

First, as pointed out in Sect. 4.6, if I is an SIN ideal, then I^* is an ultrafilter. But this is not true for NIS ideals, and we have no characterization of NIS ideals. Thus we ask the following.

Question 5.6.1. Is every weakly selective ideal on ω an NIS ideal?

We conclude with the following.

Question 5.6.2. Is it true that for every graph V on ω and every ideal I on ω, either there is an I^*-predictor for V or an I^+-predictor for V^c?

Chapter 6
Galvin's Setting: Neutral and Anonymous Predictors

6.1 Background

The result of Galvin that underlies the considerations of the present chapter first appeared in 1965 in the problem section of the *American Mathematical Monthly* (see [Gal65] and [Tho67]). It reappears in [GP76] as Lemma 1, where it is shown to have important consequences in infinitary combinatorics, set theory, and logic. We restate it here as a hat problem.

Theorem 6.1.1 (Galvin). *Consider the situation in which the set A of agents is the set of natural numbers, the set K of colors is arbitrary, and every agent sees the hats of higher numbered agents but—and this is important—the agent doesn't know where in line he stands. Then there exists a finite-error predictor.*

More precisely, Galvin's theorem asserts the existence, for an arbitrary set K, of a function $p : {}^{\omega}K \to K$ such that $x_n = p(\langle x_{n+1}, x_{n+2}, \ldots \rangle)$ for all but finitely many n. In particular, each strategy depends only on the tail segment being viewed, and does not depend on which agent is viewing it.

It turns out that, in some respects, the μ-predictor has its roots in a 40-year-old failed attempt to prove Galvin's theorem. D. L. Silverman [Sil66] proposed the following strategy: Well order the hat colorings, and for a given sequence $s \in {}^{\omega}K$ find the first hat coloring $\langle x_0, x_1, \ldots \rangle$ in the ordering of which s is a proper end segment $\langle x_{n+1}, x_{n+2}, \ldots \rangle$. Set $p(s) = x_n$.

The failure of this proof was pointed out by B. L. D. Thorp [Tho67]: If $\langle 1, 1, 0, 1, 0, 1, 0, \ldots \rangle$ is the first hat coloring in the well-ordering, then the sequence $s = \langle 1, 0, 1, 0, 1, 0, \ldots \rangle$ will elicit a guess of 1 which will be wrong infinitely often.

Nevertheless, the μ-predictor can be made to work in this context in a couple of different ways. The first way is to handle the case of eventually periodic hat colorings separately; this is easy and doesn't require the axiom of choice. The μ-predictor can then be invoked on the remaining class of hat colorings. The second way, which we use here, is to choose a well-ordering of hat colorings in which the

C.S. Hardin and A.D. Taylor, *The Mathematics of Coordinated Inference*, Developments in Mathematics 33, DOI 10.1007/978-3-319-01333-6__6, © Springer International Publishing Switzerland 2013

periodic colorings come first. The basic idea of the following proof will be extended later in this chapter to yield stronger results, such as Theorem 6.5.2.

Proof. Let $\preceq$ be a well-ordering of $^\omega K$ in which the periodic colorings appear first. For $x \in {}^\omega K$ and $n \in \omega$, let $x{\downarrow}n = \langle x_0, x_1, \ldots \rangle {\downarrow} n = \langle x_{n+1}, x_{n+2}, \ldots \rangle$. Define $p : {}^\omega K \to K$ as above: Given $s \in {}^\omega K \to K$, find the $\preceq$-least $x \in {}^\omega K$ such that $s = x{\downarrow}n$ for some n, and let $p(s) = x_n$. (If there is more than one n such that $s = x{\downarrow}n$, take the least one.) Define the predictor P by $P(f)(a) = p(f{\downarrow}a)$.

Take any coloring $f \in {}^\omega K$. If f is eventually periodic then, because periodic functions appear first in $\preceq$, all agents in the periodic tail will guess according to the periodic pattern, so their guesses are all correct. Now suppose f is not eventually periodic. Let $x \in {}^\omega K$ be $\preceq$-minimal such that $f{\downarrow}m = x{\downarrow}n$ for some $m, n \in \omega$. Note that for distinct $r, r' \in \omega$, we must have $x{\downarrow}r \neq x{\downarrow}r'$; otherwise, r, r' would witness that x is eventually periodic, which in turn would imply that f is eventually periodic. In particular, in the following step, we do not have to worry about non-uniqueness in aligning tails of f with tails of x. For any $a > m$, we have $f{\downarrow}a = x{\downarrow}(n + a - m)$, and agent a will guess $p(f{\downarrow}a) = x_{n+a-m} = f_{m+a-m} = f_a$, so a guesses correctly. $\square$

If the set of agents is the set of natural numbers, then Galvin's theorem is stronger than the Gabay-O'Connor theorem. But the Gabay-O'Connor theorem generalizes in ways that Galvin's theorem appears not to. For example, we showed in Chap. 4 that if the agent set is an ordinal α and each agent sees the hats of all higher-numbered agents, then—in the Gabay-O'Connor context in which each agent knows where in "line" he stands—there is a finite-error predictor. In Galvin's context, however, this turns out to work, as we now show, iff $\alpha < \omega^2$.

Theorem 6.1.2. *Assume that the set A of agents is an ordinal α and that each agent can see the hats of all the higher-numbered agents, but that the agents don't know where in line they are standing. Then, for every set of two or more colors, there is a finite-error predictor iff $\alpha < \omega^2$.*

Proof. If $\alpha < \omega^2$, then, for any set K of colors, the agents in each of the finitely many ω-blocks can use the strategy from Galvin's theorem among themselves. If $\alpha = \omega^2$, then consider the hat coloring in which the successor ordinals have red hats and the limit ordinals have blue hats. Note that each agent sees the same tail, and so any predictor of Galvin's type immediately yields infinitely many incorrect guesses.

Conceivably, different agents could have different strategies, despite not knowing where in line they are. To handle this extra generality, while not having agents know where they are standing, we must adopt something like the following formalization: In addition to the coloring $f \in {}^\alpha K$, we have a bijection $\pi : A \to \alpha$ that assigns each agent to a position in line. (So, while Alice and Bob might have different strategies, their positions in line are not fixed in advance, but are instead determined by π.) Each agent a, ignorant of $\pi(a)$, must individually use a strategy p_a of Galvin's type. With a predictor of this more general type, under the same hat coloring as above, if only finitely many guess red when seeing this tail (or only finitely many guess blue), then infinitely many of them will be wrong, regardless of how they are arranged in line. If infinitely many guess red and infinitely many guess blue when seeing this tail, then they will all guess incorrectly if π places the blue-guessers on successor ordinals and the red-guessers on limit ordinals.   □

6.2   Applications to Logic and Set Theory

Galvin's original interest in Theorem 6.1.1 was inspired by some model-theoretic considerations that we now describe. Leon Henkin had introduced in [Hen59] formulas with non-well-ordered quantifiers. In particular, he considered sentences such as

$$\cdots \exists y_1 \forall x_1 \exists y_0 \forall x_0 [x_0 = y_0 \vee x_1 = y_1 \vee \cdots] \tag{6.1}$$

which is, by definition, equivalent to the second-order sentence

$$\exists f_0 f_1 f_2 \cdots \forall x_0 x_1 x_2 \cdots [x_0 = f_0(x_1, x_2, \ldots) \vee x_1 = f_1(x_2, x_3, \ldots) \vee \cdots]. \tag{6.2}$$

Theorem 6.1.1 shows that this sentence is true in every domain K and that, in fact, the functions $f_0, f_1, \cdots$ can all be taken to be the finite-error predictor in Theorem 6.1.1.

However, the "formal negation" of (6.1) is the sentence

$$\cdots \forall y_1 \exists x_1 \forall y_0 \exists x_0 [x_0 \neq y_0 \wedge x_1 \neq y_1 \wedge \cdots] \tag{6.3}$$

which is, again by definition, equivalent to the second-order sentence

$$\exists f_0 f_1 f_2 \cdots \forall y_0 y_1 y_2 \cdots [f_0(y_0, y_1, \ldots) \neq y_0 \wedge f_1(y_1, y_2, \ldots) \neq y_1 \wedge \cdots] \tag{6.4}$$

and this sentence is true in any domain K with two or more elements.

As pointed out in [GP76], this paradoxical behavior of Henkin quantifiers was first shown by Jerome Malitz with a more complicated example in [Mal66].

More than a decade after Galvin first established Theorem 6.1.1, he and Karel Prikry produced a number of applications in [GP76] to Jonsson algebras and partition relations. To describe the former, we need a few definitions. An ω-ary algebra is a pair (K, P) where K is a set and P is a partial function from $^{\omega}K$ to K. If (K, P) is an ω-ary algebra and $J \subseteq K$, then $P^*(J) = \{ P(x) : x \in \text{dom}(P) \cap ^{\omega}J \}$, and J is a subalgebra of (K, P) if $P^*(J) \subseteq J$. A Jonsson algebra of cardinality λ is an ω-ary algebra of cardinality λ with no proper subalgebra of cardinality λ.

Kenneth Kunen's proof [Kun71] of the inconsistency of a nontrivial elementary embedding of the universe into itself made use of the existence of a Jonsson algebra of cardinality λ for cardinals such that $2^\lambda = \lambda^{\aleph_0}$. The existence of these algebras is a consequence of the following theorem of Paul Erdős and Andras Hajnal [EH66].

Theorem 6.2.1. *For every infinite cardinal λ, there is a Jonsson algebra of cardinality λ.*

Galvin and Prikry observed that a relatively easy argument (that we now give) shows that if P is the predictor given by Theorem 6.1.1 for $K = \lambda$, then some subalgebra A of (K, P) is a Jonsson algebra of cardinality λ. Their first observation was that for every infinite sequence $A_0 \supseteq A_1 \supseteq \cdots$ of subsets of λ, there is some n for which $A_n \subseteq A_{n+1} \cup P^*(A_{n+1})$. If not, we could choose $x_n \in A_n - (A_{n+1} \cup P^*(A_{n+1}))$, and lose no generality in assuming the x_n's are increasing. But then there is some k such that $x_k = P(\langle x_{k+1}, x_{k+2}, \ldots\rangle) \in A_{k+1} \subseteq A_k$, a contradiction. Given this observation, we can now choose $A \in [\lambda]^\lambda$ such that $A \subseteq X \cup P^*(X)$ for every $X \in [A]^\lambda$. But, in fact, for each such X we actually have that $X \subseteq P^*(X)$, because for each $a \in A$ we have $A \subseteq (X - \{a\}) \cup P^*(X - \{a\})$. It now follows that A is the desired subalgebra.

It is also worth noting that the above argument shows that for every infinite cardinal λ, we have that the relation $\lambda \to [\lambda]^\omega_\lambda$ fails. Much more along these same lines can be found in [GP76].

6.3 Neutral and Anonymous Predictors

Throughout this section, we are concerned with the special case of one-way visibility on ω and (unless otherwise specified) finitely many colors. Consider the predictor in which an agent guesses the least color seen infinitely often. This predictor lacks a certain symmetry with respect to color: if we permute the colors, the guesses are not necessarily permuted likewise. On the other hand, it does have some symmetry with respect to agents: each agent turns what they see into a guess in the same way. We say that this predictor is not *neutral*, but it is *anonymous*. We make these notions precise below, and consider the existence of neutral and anonymous predictors.

Definition 6.3.1. A predictor P for a hat problem with κ colors is *neutral* if, for every coloring f and permutation $\sigma : \kappa \to \kappa$, $P(\sigma \circ f) = \sigma \circ (P(f))$. I.e., if the colors are permuted, then all the guesses are permuted likewise.

For $n \in \omega$, define $t^n : \omega \to \omega$ by $t^n(a) = a + n$ (equivalently, let t be the successor function and let t^n be the n-fold composition of t with itself). A predictor P for a hat problem with $A = \omega$ is *anonymous* if $P(f \circ t^n) = P(f) \circ t^n$ for every $n \in \omega$ and coloring f. Though this definition of anonymity is concise and makes obvious the duality with neutrality, we find it convenient to give the following alternative (but equivalent) definition of anonymity.

Given $f \in {}^\omega\kappa$ and $n \in \omega$, define $f{\downarrow}n \in {}^\omega\kappa$ by $(f{\downarrow}n)(j) = f(n + 1 + j)$. (Equivalently, $f{\downarrow}n = f \circ t^{n+1}$.) Note that $f{\downarrow}n$ captures what agent n sees under coloring f; we call $f \downarrow n$ the *observation* of agent n under f. With one-way visibility on ω and κ colors, a predictor P is *anonymous* if there is a function $p : {}^{\omega}\kappa \to \kappa$ such that for every $f \in {}^{\omega}\kappa$ and every $n \in \omega$, $P(f)(n) = p(f \downarrow n)$. (So, each agent is using the same function p to map observations to guesses.) We call p the *common strategy* of P.

Certainly, the existence of a non-principal ultrafilter on ω is sufficient to get a neutral minimal predictor for any finite set of colors: the agents guess the color that they see occurring on a measure one set. However, the existence of such an ultrafilter is not provable in ZF (and the predictor just described is not anonymous), so the following theorem requires a more subtle argument.

Theorem 6.3.2 (ZF). *With full visibility to the right on ω and finitely many colors, there exists a neutral and anonymous predictor that ensures at least one correct guess.*

Assume there are k colors. The key idea in the proof below is that we assign types to agents based on the order in which colors first appear to them. Agents then focus their attention on higher-numbered agents of the same type when producing their guesses.

Definition 6.3.3. For a given $f : \omega \to k$, define $\tau_f \in {}^{<\omega}k$ by letting $\tau_f(0) = f(0)$ be the first value occurring in f, $\tau_f(1)$ be the next *distinct* value occurring in f, and so on. We call τ_f the *color type of f*. A color type τ induces an ordering $\leq_\tau$ of its entries in the obvious way; formally, $a \leq_\tau b \iff \tau^{-1}(a) \leq \tau^{-1}(b)$. We assign color types to agents according to what they see: for an agent n, under coloring f, the *color type of n (under f)* is $\tau_{f \downarrow n}$. The *canonization* of f is $\lfloor f \rfloor = \tau_f^{-1} \circ f$ (which is well-defined since τ_f is one-to-one with range containing the range of f). For example, with $k = 3$, if $f = (c, c, a, b, c, \ldots)$, we have $\tau_f = (c, a, b)$, and $\lfloor f \rfloor = (0, 0, 1, 2, 0, \ldots)$.

Proof of Theorem 6.3.2. We define our neutral, anonymous predictor P in terms of its common strategy $p : {}^{\omega}k \to k$. Given $f \in {}^{\omega}k$, let $\tau = \tau_f$, let $M = \{m \in \omega : \tau_{f \downarrow m} = \tau\}$, and let c be the $<_\tau$-least color occurring infinitely often in $f|M$; we define $p(f) = c$. We can restate P informally as follows: each agent determines own his color type τ, looks at the agents later in line who share this color type, and guesses the $<_\tau$-least color that occurs infinitely often among *just those agents*. Note that P is anonymous and neutral.

To verify that P is a minimal predictor, take any $f \in {}^{\omega}k$. Since there are only finitely many possible types, there must be an infinite set S of agents who share the same color type τ. Let c be the $<_\tau$-least color occurring infinitely often in $f|S$. Then all agents in S guess c and infinitely many will be correct. $\qquad\square$

Assuming the consistency of a large cardinal, Theorem 3.5.3 shows that we cannot prove in ZF + DC that a minimal predictor exists for ω colors. If neutrality is required, we can go further and prove the non-existence of a minimal predictor for ω colors.

Theorem 6.3.4 (ZF). *With full visibility to the right on ω and ω colors, there is no neutral predictor that ensures even one correct guess.*

Proof. Observe that, under a neutral predictor, if there are two or more colors that an agent n does not see, then n must guess a color among those visible to n: if agent n guesses some color k not visible to n, we can let σ be a permutation that fixes the colors visible to n while swapping k with another color, and σ would contradict the predictor's neutrality.

Let P be any neutral predictor for ω colors. Define $f : \omega \to \omega$ by $f(n) = 2n$. Then, by the above observation, every agent must guess incorrectly, since no agent's color is among the colors visible to that agent.   □

6.4   Neutralizing Predictors

We now look at a way to produce a neutral predictor from a given predictor. Given a predictor P, we define the *basic neutralization B* of P by[1]

$$B(f)(n) = \tau_{f\downarrow n}(P(\tau_{f\downarrow n}^{-1} \circ f)(n)).$$

In the special case where P is anonymous with common strategy p, this can be expressed more succinctly by $B(f)(n) = \tau_{f\downarrow n}(p(\lfloor f\downarrow n\rfloor))$. In either case, to produce n's guess, we canonize (from n's perspective), apply the original predictor, and then apply the inverse of the permutation used to canonize. Note that, unlike the proof of Theorem 6.3.2, we do not restrict our attention to agents of the same type (but that would also work in Theorem 6.4.2 below, provided P is anonymous; the relevance of anonymity when restricting attention to agents of the same type is that, since agents do not know how many agents before them share the same type, they are unsure of their position among agents of the same type).

Proposition 6.4.1. *The basic neutralization B of a predictor P is neutral. Furthermore, if P is anonymous then so is B.*

The basic neutralization of a minimal predictor is not necessarily a minimal predictor. For example, if P is the predictor that guesses the least color seen infinitely often, then B will guess incorrectly everywhere under coloring $(0, 1, 0, 1, \ldots)$. The basic neutralization is better suited for I^*-predictors (in particular, for finite-error predictors and $\mathcal{U}$-predictors), as the following theorem shows.

Theorem 6.4.2. *Suppose I is an ideal on ω and that P is an I^*-predictor for k colors. Let B be the basic neutralization of P. Then B is a neutral I^*-predictor.*

---

[1] There is a slight abuse of notation here: $\tau_{f\downarrow n}^{-1} \circ f$ might not be defined for $k \leq n$, since color $f(k)$ might not occur in $f\downarrow n$; however, that is not a significant issue, since a predictor's guess at n must ignore colors of agents $k \leq n$ anyhow.

Proof. Take any coloring $f \in {}^{\omega}k$. Let $T = \{\tau_{f\downarrow n} : n \in \omega\}$, the set of color types of agents under f; note that T is finite. For $\tau \in T$, let $A_\tau = \{n \in \omega : \tau_{f\downarrow n} = \tau\}$.

Take any $\tau \in T$. $P(\tau^{-1} \circ f)\Delta(\tau^{-1} \circ f) \in I$, so $(\tau \circ P(\tau^{-1} \circ f))\Delta(\tau \circ \tau^{-1} \circ f) \in I$, so

$$(\tau \circ P(\tau^{-1} \circ f))\Delta f \in I. \tag{6.5}$$

(Again, there is a slight abuse of notation here, since $\tau^{-1} \circ f$ might not be total, but this is not a significant issue.) The behavior of B for agents in A_τ is to guess according to $\tau \circ P(\tau^{-1} \circ f)$, so by (6.5), the set of agents in A_τ who guess incorrectly under predictor B and coloring f is in I. Since there are only finitely many color types τ, it follows that the set of agents who guess incorrectly under predictor B and coloring f is in I. $\square$

6.5 Combining with Robustness

Recall that a *robust* predictor is one which respects $=^*$ (that is, agents ignore finite differences). The basic neutralization above typically breaks robustness, since color types can change when a single agent's color is changed. However, if we have a non-principal ultrafilter $\mathcal{U}$ on ω, we can neutralize a predictor P in a different way that does preserve robustness (but which can break anonymity): where agents used their own color types above, we instead have every agent use the unique color type that is shared by a measure one set of agents.

In ZFC, we can get robust anonymous neutral finite-error predictors for k colors.

Theorem 6.5.1. *For any finite k, there exists a robust anonymous neutral finite-error predictor for k colors.*

Proof. Say a coloring f is *tail-like* if any color occurring at all in f occurs cofinally often; note that periodic colorings are tail-like. Let $\preceq$ be any well-ordering of the colorings in which the periodic colorings appear first, followed by the other tail-like colorings, followed by the remaining colorings. Each agent n guesses as follows under coloring f. Let g_n be $\preceq$-minimal such that there exist m_n and color type τ_n such that $f \downarrow i =^* \tau_n \circ g_n \downarrow m_n$; fix a minimal such m_n, and the unique τ_n for g_n and m_n that involves no more colors than needed. Note that g_n is always tail-like because of the use of $=^*$ and our choice of $\preceq$. Let $P(f)(n) = \tau(g_n(m))$.

Note that P is anonymous, neutral, and robust. It remains to be shown that P is a finite-error predictor. Take any coloring f. Fix N such that g_N is $\preceq$-minimal among $\{g_n : n \in \omega\}$, and let $g = g_N$. Then $g_k = g$ for all $n \geq N$. If f is eventually periodic, then g will be periodic, the agents will guess according to the periodic pattern, and all of them will be correct in the tail where f is periodic.

If f is not eventually periodic, we claim that g has the following non-self-similarity: there do not exist distinct m, m' with types τ, τ' (not necessarily distinct) such that $\tau \circ g \downarrow m =^* \tau' \circ g \downarrow m'$. Suppose for a contradiction that such m, m', τ, τ'

exist, and assume without loss of generality that $m < m'$; let $q = m' - m$. By enlarging m and m' as necessary, we can also assume without loss of generality that $\tau \circ g{\downarrow}m = \tau' \circ g{\downarrow}m'$. Let $\sigma = \tau^{-1} \circ \tau'$. Let $h = g{\downarrow}m = g \circ t^{m+1}$ where t is the successor function. Then

$$\sigma \circ h \circ t^q = \tau^{-1} \circ \tau' \circ g \circ t^{m+1} \circ t^q$$

$$= \tau^{-1} \circ \tau' \circ g \circ t^{m'+1}$$

$$= \tau^{-1} \circ \tau' \circ g{\downarrow}m'$$

$$= \tau^{-1} \circ \tau \circ g{\downarrow}m$$

$$= g{\downarrow}m = h.$$

In particular, letting $o = o(\sigma)$ (the order of σ as a permutation of the colors appearing in g), $h = \sigma^o \circ h \circ (t^q)^o = h \circ t^{qo}$, so h is periodic. It follows that g and f are eventually periodic, a contradiction.

With this lack of self-similarity in g, the agents $n \geq N$ guess according to g in a consistent way: if $n, n' \geq N$, then $m_{n'} - m_n = n' - n$, and $\tau_{n'} = \tau_n$. It follows that all but finitely many of them guess correctly.   □

If we omit neutrality in the above result, we can make it work for an arbitrary set of colors, yielding the result below. (Theorem 6.3.4 tells us that the loss of neutrality is necessary.)

Theorem 6.5.2. *For any cardinal κ, there exists a robust anonymous finite-error predictor for κ colors.*

Proof. The proof is the same as for the previous theorem, except that we disregard the notion of tail-like, and remove the apparatus for neutrality (that is, replace $\tau, \tau', \tau_n, \sigma$ with the identity).   □

6.6   Robust Neutral Predictors and the Axiom of Choice

Combining robustness and neutrality is "hard" in the sense that ZF + DC cannot prove the existence of a robust neutral predictor for two colors (even if we do not require any correct guesses). Our approach here has much in common with Sect. 3.4, in that we use the existence of such a predictor to contradict the assumption— denoted BP in Sect. 3.4—that all sets of reals have the property of Baire.

Lemma 6.6.1 (ZF). *Suppose there exists a function $p : {}^\omega 2 \to 2$ such that p respects $=^*$ (robustness) and if τ is a permutation of 2, $(\tau \circ p)(x) = p(\tau \circ x)$ for all $x \in {}^\omega 2$ (neutrality). Then $\neg$BP.*

Proof. Let $\mathcal{C} = {}^{\omega}2$, treated as the Cantor set (with the usual topology). For $i = 0, 1$, let $A_i = \{ x \in \mathcal{C} : p(x) = i \}$. Then A_0 and A_1 form a partition of $\mathcal{C}$. Define $c : \mathcal{C} \to \mathcal{C}$ by $c(x)(i) = 1 - x(i)$ (considering $\mathcal{C}$ as a subset of $[0, 1]$ in the usual way, this happens to be the same as $c(x) = 1 - x$). The function c is an automorphism of $\mathcal{C}$ that maps A_0, A_1 to each other. In particular, if A_i is meager (as a subset of $\mathcal{C}$), then so is A_{1-i}, making $\mathcal{C} = A_0 \cup A_1$ meager, a contradiction; so, neither A_i is meager.

Suppose for a contradiction that A_0 has the property of Baire. Then there is an open set U_0 such that $A_0 \Delta U_0$ is meager. Since A_0 is not meager, $U_0 \neq \emptyset$, so there is some $\alpha \in {}^{<\omega}2$ such that $[\alpha] \subseteq U_0$ (where $[\alpha] = \{ f \in \mathcal{C} : \alpha \text{ is a prefix of } f \}$ is the basic open set given by α). In particular, $[\alpha] - A_0$ is meager. Let β be the complement of α (e.g., if $\alpha = (0, 0, 1)$, then $\beta = (1, 1, 0)$). Let $U_1 = c[U_0]$; we have that $A_1 \Delta U_1$ is meager, and $[\beta] \subseteq U_1$, so $[\beta] - A_1$ is meager. Let $n = |\alpha| = |\beta|$, and let $t : \mathcal{C} \to \mathcal{C}$ be the function that toggles the first n bits of a real (that is, $t(x)(i) = 1 - x(i)$ for $i < n$, and $t(x)(i) = x(i)$ for $i \geq n$). Then t is an automorphism of $\mathcal{C}$ that sends $[\alpha]$ and $[\beta]$ to each other. However, by the robustness of p, τ does not alter membership in A_0 (or A_1). In particular, $\tau[[\alpha] - A_0] = [\beta] - A_0$ is meager.

Since $[\beta] = ([\beta] - A_0) \cup ([\beta] - A_1)$, a union of two meager sets, $[\beta]$ is meager, a contradiction. So, A_0 does not have the property of Baire.   □

Theorem 6.6.2. *Assuming* ZF *is consistent,* ZF $+$ DC *cannot prove the existence of a robust neutral predictor for two colors on* ω.

Proof. For such a predictor P, applying the previous lemma to the individual strategy p of agent 0 ($p(f) = (P(f))(0)$) yields $\neg$BP. However, $\neg$BP is not provable from ZF $+$ DC (assuming ZF is consistent).   □

Chapter 7
The Topological Setting

7.1 Background

A topological space is T_0 if for every two distinct points in the space, there is a neighborhood of one not containing the other. Equivalently, it is T_0 if distinct points have distinct neighborhood systems. A space is T_1 if for every pair of distinct points each has a neighborhood not containing the other. And although of little relevance for what is to follow, a space is T_2 or *Hausdorff* if every pair of distinct points have neighborhoods that are disjoint.

Among the topological spaces that we will be interested in are the ones arising from a partial ordering by either declaring a set to be open if it is closed upward in the ordering (the *upward topology*) or closed downward in the ordering (the *downward topology*). For example, the interval $(-\infty, 0]$ is open in the downward topology on the reals. These topologies are not typically T_1, but they are always T_0.

We will be asserting that certain things happen except on a set that is "topologically small," and we want to do so in a way that generalizes a couple of particular results. Thus, on the real line $\mathbf{R}$ with the downward topology, we want these small sets to be the well-ordered subsets of $\mathbf{R}$, and for the upward topology on an ordinal, we want these small sets to be the finite sets. The following well-known notions achieve both.

Definition 7.1.1. A set S in a topological space X is *weakly scattered* if for every nonempty $T \subseteq S$ there exists some $x \in T$ and some neighborhood N of x such that $N \cap T$ is finite. We call such points *weakly isolated points of T*. The set S is *scattered* if the conclusion can be strengthened to $N \cap T = \{x\}$, in which case these are called *isolated points of T*.

What we are calling "weakly scattered" is called "separated" by Morgan [Mor90]; he attributes the definition to Cantor. Every scattered set is weakly scattered, and it is straightforward to show that the two notions are equivalent in T_0 spaces. The concept of a weakly scattered set is a smallness notion in the sense that this class is closed under the formation of subsets and finite unions.

C.S. Hardin and A.D. Taylor, *The Mathematics of Coordinated Inference*, Developments in Mathematics 33, DOI 10.1007/978-3-319-01333-6_7, © Springer International Publishing Switzerland 2013

There is a game-theoretic characterization of weakly scattered sets in [HT09], special cases of which occurred in [Fre90], that goes as follows. Given a space X and set $S \subseteq X$, Players I and II take turns, with Player I choosing elements of S and Player II choosing open sets. Player I must choose his point in the last open set chosen by Player II and Player II must choose his open set to be a neighborhood of Player I's last chosen point. Player I wins iff all his choices are distinct. The set S is weakly scattered iff Player II has a winning strategy, and not weakly scattered iff Player I has a winning strategy. This characterization generalizes the fact that a set is well ordered iff it has no infinite descending chains.

With the downward topology on any partial order, the scattered sets are the well-founded subsets (so with the upward topology, the scattered sets are the co-well-founded subsets). In particular, with the downward topology on $\mathbf{R}$, the scattered sets are the well-ordered subsets, and with the upward topology on any ordinal, the scattered sets are the finite subsets. The usual topology of the reals is not of much interest to us here, but the scattered sets there are countable and nowhere dense; in fact, they are precisely the countable G_δs. This observation goes back to the early twentieth century, but for a readily accessible proof, see [DG76].

If S is a scattered set in a space X, then S gives rise to what we call the *canonical decomposition* $\langle A_\alpha : \alpha < \lambda \rangle$ *of* X *associated with* S. In this decomposition, $A_0 = X - S$ and, for $\alpha > 0$, A_α is the set of isolated points of $X - \cup\{ A_\beta : \beta < \alpha \}$. This decomposition is directly related to the Cantor-Bendixson derivative.

7.2   The Scattered Sets Result

In what follows (some of which is drawn from [HT09]) we fix a topological space X, and we assume Y is a set with two or more elements. Intuitively, X is the set of agents and Y is the set of colors.

Definition 7.2.1. For each $x \in X$ we define an equivalence relation $\approx_x$ on $^X Y$ by $f \approx_x g$ iff f and g agree on a deleted neighborhood $N - \{x\}$ of x. We let $[f]_x$ denote the equivalence class of f under $\approx_x$. Any predictor for such an equivalence relation will be called a *near neighborhood predictor*.

We will be trying to guess $f(x)$ from the values of f near, but not at, x. The equivalence class $[f]_x$ precisely describes how much information we have about f when we produce this guess. That is, our guess must be based only on $[f]_x$; if we could use f directly when guessing, we could cheat and use $f(x)$ as our guess. If P is a predictor, then a set $E \subseteq X$ is an *error set for* P if there exists a function $f : X \to Y$ such that P guesses incorrectly for f at x for every $x \in E$. Error sets are small in the sense that singleton sets are always error sets and the class of error sets is (by definition) closed under subset formation.

The following variant of the μ-predictor will be of particular interest. We actually find it convenient to throw away a certain amount of information, by disregarding finite differences.

Definition 7.2.2. For $f, g : X \to Y$ and $V \subseteq X$, we say that f and g *differ only finitely* on N if $\{x \in N : f(x) \neq g(x)\}$ is finite. For $x \in X$, we define the equivalence relation $\approx_x^*$ on $^X Y$ by $f \approx_x^* g$ iff f and g differ only finitely on some neighborhood N of x. Let $[f]_x^*$ denote the equivalence class of f under $\approx_x^*$. Fix a well-ordering $\preceq$ of $^X Y$. The μ^*-*predictor* is the predictor $\mu^* = \{\mu_x^* : x \in X\}$ defined by letting $\mu_x^*([f]_x) = \langle f \rangle_x(x)$, where $\langle f \rangle_x$ is the $\preceq$-least element of $[f]_x^*$.

In other words, at any point x, the μ^*-predictor guesses according to the $\preceq$-least element of $^X Y$ that, overlooking finite differences, is consistent with the information available.

Our main results are the following. The first shows that the μ^*-predictor guesses correctly except on weakly scattered sets of points, and the second shows that one cannot improve on this for T_0 spaces.

Theorem 7.2.3. *For every $f : X \to Y$, the μ^*-predictor guesses incorrectly for f only at a set of points that is weakly scattered. If X is T_0, then this error set is scattered.*

Proof. Take any $f : X \to Y$ and let S be the set of points where the μ^*-predictor guesses incorrectly for f. Suppose $T \subseteq S$ is nonempty. We will exhibit $x \in T$ and a neighborhood N of x such that $N \cap T$ is finite. Choose $x \in T$ such that $\langle f \rangle_x$ is the $\preceq$-least element of $\{\langle f \rangle_y : y \in T\}$, and let N be a neighborhood of x such that $\{y \in N : f(y) \neq \langle f \rangle_x(y)\}$ is finite. Take any $y \in N \cap T$. Since N is a neighborhood of y, we have $\langle f \rangle_x \approx_y^* f$, so $\langle f \rangle_y \preceq \langle f \rangle_x$. Then $\langle f \rangle_y = \langle f \rangle_x$, by the minimality of $\langle f \rangle_x$. Since $y \in S$, $f(y) \neq \langle f \rangle_y(y) = \langle f \rangle_x(y)$. Therefore, $f(y) \neq \langle f \rangle_x(y)$ for all $y \in N \cap T$. It follows that $N \cap T$ is finite, since $\{y \in N : f(y) \neq \langle f \rangle_x(y)\}$ is finite. Therefore S is weakly scattered.

When X is T_0, the notions of scattered and weakly scattered coincide, so S is scattered. $\square$

Corollary 7.2.4. *For any space X of agents and set Y of colors, there exists a weakly scattered–error near neighborhood predictor.*

The following theorem is the analogue of Proposition 1.3.2 in the topological context.

Theorem 7.2.5. *For every near neighborhood predictor P and every scattered set $S \subseteq X$, there exists some $f : X \to Y$ such that P guesses incorrectly for f at every $x \in S$.*

Proof. Suppose that $P = \langle G_x : x \in X \rangle$ is a predictor and that $S \subseteq X$ is scattered. Let $\langle A_\alpha : \alpha < \lambda \rangle$ be the canonical decomposition of X associated with S. So $A_0 = X - S$ and for $\alpha > 0$ and $x \in A_\alpha$ there exists a neighborhood N of x such that $N \cap \bigcup \{A_\beta : \beta \geq \alpha\} = \{x\}$. Let f be defined arbitrarily on A_0. For $\alpha > 0$, assuming f has been defined on A_β for each $\beta < \alpha$, we define f on A_α as follows. Given $x \in A_\alpha$, let N be a neighborhood of x such that $N \cap \bigcup \{A_\beta : \beta \geq \alpha\} = \{x\}$.

Then f has already been defined on $N - \{x\}$, so $[f]_x$ has been determined. We can now define $f(x) \in Y$ so that $f(x) \neq G_x([f]_x)$, ensuring that P guesses incorrectly for f at x. □

There is a version of Theorem 7.2.5 for the case in which the space is not T_0. However, it turns out to be somewhat of a disjoint union of what we have done here and the finite case in [HT10].

In fact, the μ-predictor works unmodified in T_0 spaces, as we now show, but the proof is significantly longer. Essentially, the μ^*-predictor, by ignoring finite sets, is voluntarily refining the topology to be T_1, and this stronger separation makes the proof go more smoothly. In fact, if we stated the following theorem for T_1 spaces instead of T_0 spaces, the proof would be more akin to the proof of Theorem 7.2.3.

Theorem 7.2.6. *If X is T_0, then for every $f : X \to Y$, the μ-predictor is a near neighborhood predictor that guesses incorrectly for f only at a set of points that is scattered.*

Proof. Take any $f : X \to Y$ and let S be the set of points where the μ-predictor (denoted just μ in what follows) guesses incorrectly for f. Define $\leq$ on X by $x \leq y$ if every neighborhood of y is also a neighborhood of x. We claim that for $x, y \in S$, $x < y \Rightarrow \langle f \rangle_x \prec \langle f \rangle_y$. Suppose $x, y \in S$ with $x < y$. Let V be a neighborhood of y (and hence also of x) witnessing $f \approx_y \langle f \rangle_y$. Since $x \not\geq y$, x has a neighborhood U with $y \notin U$. Then f and $\langle f \rangle_y$ agree on $U \cap V$, witnessing $f \approx_x \langle f \rangle_y$, so $\langle f \rangle_x \preceq \langle f \rangle_y$. However, $\langle f \rangle_y(x) = f(x)$ (since $x \in V - \{y\}$) and $f(x) \neq \langle f \rangle_x(x)$ (since $x \in S$), so $\langle f \rangle_x \neq \langle f \rangle_y$. We now have $\langle f \rangle_x \prec \langle f \rangle_y$, establishing the claim.

By the claim, it follows that S is well-founded in $\leq$; otherwise, it would induce an infinite descending chain in $\preceq$.

Suppose for a contradiction that S is not scattered. Then, since X is T_0, S is not weakly scattered, so there exists a nonempty $S' \subseteq S$ with no weakly isolated points. Note that any nonempty intersection of an open set with S' is infinite with no weakly isolated points.

Let $x \in S'$ be such that $\langle f \rangle_x$ is $\preceq$-minimal among $\{ \langle f \rangle_y : y \in S' \}$, and let V be a neighborhood of x witnessing $f \approx_x \langle f \rangle_x$. Let $S'' = S' \cap V - \{x\}$. Then S'' is infinite.

Take any $y \in S''$ and suppose $y \not\geq x$. Then y has a neighborhood U excluding x. Then $f \approx_y \langle f \rangle_x$, since $\langle f \rangle_x$ and f agree on $V - \{x\} \supseteq V \cap U$. It follows that $\langle f \rangle_y \preceq \langle f \rangle_x$, and hence $\langle f \rangle_y = \langle f \rangle_x$ by the minimality of $\langle f \rangle_x$. It follows that μ guesses correctly at y, since $\langle f \rangle_x$ and f agree on $V - \{x\}$, a contradiction. So, for all $y \in S''$, $y > x$.

Let Z be the set of $\leq$-minimal elements of S''. Since $\leq$ is well founded on S, every $y \in S''$ has a $z \in Z$ with $z \leq y$. Note that Z forms an antichain in $\leq$.

Take any neighborhood V' of x. We claim that $V' \cap Z$ is infinite. If not, let $V' \cap Z = \{z_1, \ldots, z_n\}$. For $1 \leq i \leq n$, we have $x < z_i$, so there is a neighborhood V_i of x that excludes z_i. Then the set $T = (\cap_i V_i) \cap V \cap V'$ is a neighborhood of x disjoint from Z. Since the complement of T is closed and hence closed upwards under $\leq$, T is in fact disjoint from S''. So, $T \cap S' = \{x\}$, contradicting the fact that S' has no weakly isolated points. Therefore, every neighborhood of x has infinite intersection with Z.

For any $z \in Z$ and a neighborhood U of z, U is a neighborhood of x (since $x < z$), so U intersects Z infinitely as shown above. So Z has no weakly isolated points.

Take $z_0 \in Z$ such that $\langle f \rangle_{z_0}$ is $\preceq$-minimal among $\{ \langle f \rangle_z : z \in Z \}$. Let U_0 be a neighborhood of z_0 witnessing $f \approx_z \langle f \rangle_{z_0}$. Take any $z_1 \in Z \cap U_0 - \{z_0\}$. Since Z is an antichain in $\leq$, $z_0 \not\leq z_1$, so z_1 has a neighborhood U_1 excluding z_0. Then f and $\langle f \rangle_{z_0}$ agree on $U_0 \cap U_1$, so $f \approx_{z_1} \langle f \rangle_{z_0}$, yielding $\langle f \rangle_{z_1} \preceq \langle f \rangle_{z_0}$. It follows that $\langle f \rangle_{z_1} = \langle f \rangle_{z_0}$ by the minimality of $\langle f \rangle_{z_0}$. So, since f and $\langle f \rangle_{z_0}$ agree on $U_0 - \{z_0\}$, μ guesses correctly at z_1, a contradiction. $\qquad \square$

7.3 Corollaries

The easy generalization of the Gabay-O'Connor theorem to the context of ideals that is described at the beginning of Sect. 4.6 follows from Theorem 7.2.6 by considering the topology on A in which each set of I-measure one containing x is a basic neighborhood of x. This topology is T_0 because singletons are in I. It is easy to see that the scattered sets are precisely the sets of I-measure zero, and thus the result follows.

The second result we derive concerns the extent to which "the present can be predicted based on the past." Here, the exact characterization of the error sets occurs in Theorems 3.1 and 3.5 in [HT08b].

Theorem 7.3.1. *There exists a predictor that will, for each function $f : \mathbf{R} \to Y$, correctly guess the value of $f(x)$ from the values of f on $(-\infty, x)$, for all x except those in a well-ordered subset of $\mathbf{R}$.*

The derivation of this from our main result uses the downward topology on $\mathbf{R}$ in which, as we mentioned, the scattered sets are the well-ordered subsets of the reals.

In point of fact, if we are trying to guess the value of $f(x)$ for some function $f : \mathbf{R} \to Y$, we don't need to know the function values on all of $(-\infty, x)$. Metaphorically speaking, the following asserts that the present can predicted from an "infinitesimal" piece of the past.

Theorem 7.3.2. *There exists a predictor that will, for each function $f : \mathbf{R} \to Y$, correctly guess the value of $f(x)$—based on the equivalence relation $f \approx_x g$ if f and g agree on (w, x) for some $w < x$—for all x except those in a countable, nowhere dense subset of $\mathbf{R}$.*

The derivation of this uses the topology on $\mathbf{R}$ in which the basic open sets are half-open intervals $(w, x]$ (so $f \approx_x g$ if f and g agree on (w, x) for some $w < x$). It is known that the scattered sets here are countable and nowhere dense. The exact characterization of the error sets in this example (as scattered sets) was absent in [HT08b].

The next two results are from [HT08b]. They use the fact that in the upward topology on a partial order, the scattered sets are the co-well-founded sets; in the special case of the usual ordering on an ordinal, these are the finite sets.

Theorem 7.3.3. *If V is a strict partial order, then under the μ-predictor, the set of agents who guess incorrectly is co-well-founded.*

Theorem 7.3.4. *Consider the hat problem in which the set of agents is an ordinal α, and each agent sees the hats of all higher-numbered agents. Then there exists a predictor ensuring that the set of agents incorrectly guessing their hat color is finite.*

Finally, suppose that we have a graph on ω that is transitive and consider the topology on ω in which the only basic neighborhood of k is the set $N_k = \{k\} \cup \{m : k < m$ and k is adjacent to $m\}$. Treating this as a hat problem, each agent k can see the hats of agents in $N_k - \{k\}$. In this context we showed in Chap. 4 that there exists a finite-error predictor iff the graph contains no infinite independent set. But what if the graph does contain an infinite independent set? It is easy to see that the scattered sets in this topology (which is T_0) are precisely the ones that contain no infinite complete subgraph, so our main results give us the following.

Theorem 7.3.5. *Consider the hat problem on ω in which each agent sees some of the hats to his right and the visibility graph is transitive. Then there is a predictor in which the error set E will never contain an infinite subset W such that, for all $m, n \in W$ with $m < n$, m can see n. Furthermore, for any predictor and any set E containing no infinite W as above, there is a way to color the hats so that the error set contains E.*

7.4   Guessing the Future

In the contexts of Theorems 7.3.1 and 7.3.2, we could make the agents' job harder by asking that they not just guess the "present," but also a bit of the "future." That is, agent t must provide a guess not just for $f(t)$, but for $f\,|[t, \infty)$, and we will call that guess correct if there is a $u > t$ such that the guess is accurate on $[t, u)$. (So t must guess the present and a bit of the future.) In point of fact, this strengthening of Theorem 7.3.2 appears in [HT08b], but it can be generalized to our topological framework as follows.

Consider the situation in which we have two different topologies, $\mathcal{T}$ and $\mathcal{T}'$, on X. The topology $\mathcal{T}$, as previously, is our *notion of visibility*: agent x's guess must be a function of the equivalence class with respect to $\mathcal{T}$ of f at x. The topology $\mathcal{T}'$ (our *notion of correctness*) describes what the agents must guess: agent x must produce a guess for the equivalence class with respect to $\mathcal{T}'$ of f at x.

For example, for Theorem 7.3.1, we would let $\mathcal{T}$ be the downward topology on **R**, and for Theorem 7.3.2, we would let $\mathcal{T}$ be the topology whose basic open sets are $(w, x]$. If we want agents to have to guess a bit of the future, we can let $\mathcal{T}'$ be the topology whose basic open sets are $[x, y)$. (Or, though it would defeat the purpose of introducing $\mathcal{T}'$ in the first place, to ask that x just guesses $f(x)$ correctly, we can take $\mathcal{T}'$ to be the discrete topology.)

Theorem 7.4.1. *Let $\mathcal{T}$, $\mathcal{T}'$ be topologies on the set X of players, such that for every $\mathcal{T}$-neighborhood A of x, there is a subneighbhorhood A' such that for all $y \in A' - \{x\}$, there is a $\mathcal{T}'$-neighborhood of y contained in $A - \{x\}$. Then, using $\mathcal{T}$ as our notion of visibility and $\mathcal{T}'$ as our notion of correctness, the set of wrong players under the μ^*-strategy is weakly scattered with respect to $\mathcal{T}$.*

Proof. Suppose there is a nonempty set W of wrong players with no points that are weakly isolated with respect to $\mathcal{T}$. We will get a contradiction. Let $x \in W$ be such that g_x is $\preceq$-minimal. Let A be a $\mathcal{T}$-neighborhood of x such that $A - \{x\} \subseteq^* \{y : g_x(y) = v(y)\}$, where v is the true coloring. Let $A' \subseteq A$ be as in the statement of the theorem. Since x is not weakly isolated with respect to $\mathcal{T}$, there are infinitely many $y \in A' \cap W$; infinitely many of them must also be in $\{y : g_x(y) = v(y)\}$. Take any one of them. Then $g_x \approx^*_y v$, so $g_y \preceq g_x$, and by minimality of g_x, $g_y = g_x$. Then, since g_x agrees with v on a $\mathcal{T}'$-neighborhood of y, y guesses correctly, a contradiction. □

For example, consider the case where agents see the recent past and are asked to guess the present and near future: we let $\mathcal{T}$ be the topology on $X = \mathbf{R}$ with basic open sets $(w, x]$ and $\mathcal{V}$ be the topology with basic open sets $[x, y)$. We can verify that these two topologies satisfy the hypotheses of Theorem 7.4.1, so the agents will be correct except on a set that is weakly scattered in $\mathcal{T}$; these sets are known to be countable and nowhere dense.

7.5 The Philosophical Problem of Induction

There have been two articles published by philosophers that have dealt with the implications of Theorem 7.3.1 to the problem of induction that is typically identified with the philosopher David Hume (1711–1776). One of these was by Alexander George [Geo07] and the other by Alexander Paseau [Pas08]; we quote extensively from the former in what follows. In particular, the views we are presenting arise from published philosophical articles and are not necessarily shared by the present authors.

Hume argued that inductive inference cannot be justified, and his argument consisted of two halves, described by George [Geo07] as follows:

> In the first half, he [Hume] argues that no demonstration—which is to say, no mathematical or logical proof—can justify extrapolations from past observations to unobserved cases. In the second, he argues that no reasoning from experience can provide such a justification.

Philosophers have addressed the second half of Hume's argument for centuries, but little attention has been paid to the first half. George explains the reason thusly:

> This [lack of attention to the first half] is partly because Hume there offers an argument in terms of conceivability that leaves many readers unimpressed; and partly because, notwithstanding his argument's weaknesses, its conclusion, that no demonstrative proof can justify induction, appears so obviously correct in the first place.

George, however, sees Theorem 7.3.2 as something that "presents a challenge to these appearances." We will not recapitulate here all that George brings out in his discussion. But it is worth noting that he is completely familiar with the result (a complete proof of which he presents in his article) and the role of the axiom of choice.

> Thus, if our only concern was to establish that *our* inductive inferences are by and large correct, then the present proof will not help. But what if what one wanted, in the first instance, to show that the past has some rational bearing on the future (where its having precisely the bearing we think it has is of subsequent concern)? What if, moreover, in a spirit of resurgent rationalism, one wanted to establish the existence of such a bearing through a priori reasoning? That is, what if one wanted to know whether it is conceptually incoherent to imagine the future's being radically unpredictable given information about the past? On one interpretation of this question we now have an answer, and a surprising one.

Finally, for another philosophical view that is particularly entertaining, see the article [HW13] by Leon Horsten and Philip Welch.

7.6 Proximity Schemes

The reader may have observed that much of the material in this chapter can be carried out in a context somewhat more general than topological spaces: it is not open sets that we are concerned with, so much as neighborhoods, and the definition of topological space puts restrictions on what can be considered a neighborhood that, for our purposes, are unnecessary. We use topological spaces in this chapter not only because they are widely familiar, but also because when we develop some theory for these more general structures, we find that the "nice" cases are in fact instances of topological spaces.

Definition 7.6.1. A *proximity scheme* V on X assigns to each $x \in X$ a filter V_x on X. Here, a filter does not have to be proper or non-principal. A set $U \in V_x$ is called a *proximity* of x.

(If we allowed V_x to be an arbitrary subset of $\mathcal{P}(X)$ above, we would have the notion of *neighborhood frame* sometimes used in the semantics of modal logic; see [FHMV95].)

Given a set X of agents and Y a set of colors, a proximity scheme V can be used to describe visibility in the following way: given $f, g \in {}^X Y$ and $x \in X$, we say $f \approx_x g$ iff $f \sqcap g \in V_x$, where $f \sqcap g = \{x : f(x) = g(x)\}$. The fact that V_x is a filter guarantees this will be an equivalence relation.

Example 7.6.2. A topology $\mathcal{T}$ on X can induce a proximity scheme V in two natural ways. The first, denoted $V = \text{Rt}\,\mathcal{T}$, is to let V_x be the set of neighborhoods of x. The second, of more interest to us and denoted $V = \text{It}\,\mathcal{T}$, is to let V_x be the filter of sets that contain a deleted neighborhood of x; when we do this, the proximity scheme captures topological visibility as it was understood earlier in the chapter.

Example 7.6.3. A binary relation R on X also induces a proximity scheme V in a natural way: for any x, we let V_x be the principal filter generated by $R(x)$ (that is, the set of sets containing $R(x)$). If R is a visibility graph, then the resulting proximity scheme gives the same notion of visibility.

As proximity schemes are a generalization of binary relations, we can extend familiar notions for binary relations to proximity schemes. In the special case where a proximity scheme is induced by a binary relation (that is, each V_x is principal), the following notions all agree with the corresponding notions for binary relations.

Definition 7.6.4. Let V be a proximity scheme on X.

(a) V is *reflexive* if $\forall x \in X\ (X - \{x\} \notin V_x)$. Equivalently, $\forall x \in X\ \forall U \in V_x\ x \in U$.
(b) V is *irreflexive* if $\forall x \in X\ (X - \{x\} \in V_x)$. Equivalently, $\forall x \in X\ \exists U \in V_x\ x \notin U$.
(c) V is *transitive* if $\forall x \in X\ \forall C \in V_x\ \exists A \in V_x\ \forall y \in A\ C \in V_y$.

Transitivity is the property of a proximity scheme that makes the proof of Theorem 4.2.1 carry over. Note that It $\mathcal{T}$ is always irreflexive, while Rt $\mathcal{T}$ is always reflexive.

Definition 7.6.5. Let V, W be proximity schemes on X.

(a) $V \le W$ if $\forall x \in X\ (V_x \supseteq W_x)$.
(b) The proximity scheme $U = V \vee W$ is defined by $U_x = V_x \cap W_x$.
(c) The proximity scheme $U = V \wedge W$ is defined by $U_x = \{A \cap B : A \in V_x \wedge B \in W_x\}$. (Equivalently, U_x is the closure of $V_x \cup W_x$ under finite intersection.)
(d) The *composition* of W with V is the proximity scheme $U = W \circ V$ defined by letting $C \in U_x$ iff $\exists A \in V_x\ \forall y \in A\ C \in W_y$; equivalently, there exists $A \in V_x$ and sets $B_y \in W_y$ for each $y \in A$ such that $\bigcup_{y \in A} B_y \subseteq C$.
(e) The *empty proximity scheme* 0 on X is the proximity scheme induced by the empty relation.
(f) The *identity proximity scheme* I on X is the proximity scheme induced by the identity relation.

Proposition 7.6.6. *The set of all proximity schemes on X, under the ordering $\le$, forms a distributive lattice with operations $\vee$ and $\wedge$.*

Many algebraic characterizations of properties of binary relations carry over to proximity schemes, as the following proposition shows.

Proposition 7.6.7. *Let V be a proximity scheme on X.*

(a) V is reflexive iff $I \le V$.

(b) V is irreflexive iff $I \wedge V = 0$.
(c) V is transitive iff $V \circ V \leq V$.

We now examine the relationship between proximity schemes and topological spaces, culminating in Theorem 7.6.17, which shows that any transitive irreflexive proximity scheme is induced by a topology.

Definition 7.6.8. Given a proximity scheme V on X, a set $U \subseteq X$ is *open* (with respect to V) if $\forall x \in U \ U \in V_x$ (equivalently, every $x \in U$ has a proximity contained in U). We use $\mathrm{Tp}\, V$ to denote the resulting topology on X.

It is easily verified that this indeed gives a topology on X. Furthermore, if V was induced by a topology $\mathcal{T}$ on X (either by taking neighborhoods or deleted neighborhoods), then the above topology coincides with $\mathcal{T}$.

Proposition 7.6.9. *For any topology $\mathcal{T}$ on X, $\mathcal{T} = \mathrm{Tp}\,\mathrm{It}\,\mathcal{T} = \mathrm{Tp}\,\mathrm{Rt}\,\mathcal{T}$.*

Proof. If $U \in \mathcal{T}$, it is easy to get $U \in \mathrm{Tp}\,\mathrm{It}\,\mathcal{T}$ and $U \in \mathrm{Tp}\,\mathrm{Rt}\,\mathcal{T}$.

If $U \notin \mathcal{T}$, let $x \in U$ such that no open neighborhood of x is contained in $\mathcal{T}$; it follows that no deleted neighborhood of x is contained in $\mathcal{T}$ either. So, whether we consider $\mathrm{Rt}\,\mathcal{T}$ or $\mathrm{It}\,\mathcal{T}$, no proximity of x is contained in U, so U is not open in $\mathrm{Tp}\,\mathrm{It}\,\mathcal{T}$ or $\mathrm{Tp}\,\mathrm{Rt}\,\mathcal{T}$.   □

Definition 7.6.10. Say that a topology on X is *locally T_1* if every $x \in X$ has an open neighborhood U such that $U - \{x\}$ is open.

Proposition 7.6.11. *Being locally T_1 is a separation condition that lies strictly between being T_0 and being T_1. That is, the class of T_1 spaces is properly contained in the class of locally T_1 spaces, which is properly contained in the class of T_0 spaces.*

Proposition 7.6.12. *For any topology $\mathcal{T}$, $\mathrm{Rt}\,\mathcal{T}$ is transitive.*

Proposition 7.6.13. *For any topology $\mathcal{T}$ on X, if $\mathcal{T}$ is locally T_1, then $\mathrm{It}\,\mathcal{T}$ is transitive.*

Proof. Suppose $\mathcal{T}$ is locally T_1 and let $V = \mathrm{It}\,\mathcal{T}$. Take any $x \in X$ and $C \in V_x$. Then $\{x\} \cup C \in \mathcal{T}$, so there exists some open neighborhood $U' \in \mathcal{T}$ of x such that $C' - \{x\} \in \mathcal{T}$. Then $C' - \{x\}$ is open with respect to V, so every $y \in C' - \{x\}$ has a proximity contained in $C' - \{x\} \subseteq U$. Therefore, V is transitive.   □

We see below in Theorem 7.6.17 that the converse also holds. Its proof is elementary once one focuses on the notion of interior.

Definition 7.6.14. Given a proximity scheme V on X and a set $C \subseteq X$, we define the *V-interior* of C to be $C^{(V)} = \{ x \in C : C \in V_x \}$.

Lemma 7.6.15. *Suppose V is a transitive proximity scheme on X, $x \in X$, and $C \in V_x$. Then $C^{(V)} \in V_x$.*

Proof. Since V is transitive and $C \in V_x$, $\exists A \in V_x \; \forall y \in A \; C \in V_y$. Without loss of generality, $A \subseteq C$; together with $\forall y \in A \; C \in V_y$, this gives $A \subseteq C^{(V)}$, so $C^{(V)} \in V_x$. □

Lemma 7.6.16. *Suppose V is transitive and $C \subseteq X$. Then $C^{(V)}$ is open. Also, for any $x \in X$ such that $C \in V_x$, $C^{(V)} \cup \{x\}$ is also open.*

Proof. Take any $x \in C^{(V)}$. Then $C \in V_x$, so by the previous lemma, $C^{(V)} \in V_x$. Therefore, $C^{(V)}$ is open.

If $x \in X$ such that $C \in V_x$, then $C^{(V)} \in V_x$. We already have $C^{(V)} \in V_y$ for all $y \in C^{(V)}$. It easily follows that $\forall y \in C^{(V)} \cup \{x\} \; (C^{(V)} \cup \{x\} \in V_y)$, so $C^{(V)} \cup \{x\}$ is open. □

Theorem 7.6.17. *Let V be an irreflexive transitive proximity scheme on X. Then $\mathrm{Tp}\, V$ is locally T_1 and $V = \mathrm{It}\,\mathrm{Tp}\, V$.*

The intuition behind the proof is that, since the open elements of V_x are dense in V_x, we can recover V_x from topological information.

Proof. Let $\mathcal{T} = \mathrm{Tp}\, V$. Take any $x \in X$. Then $X - \{x\} \in V_x$ (since V is irreflexive). Let $U = (X - \{x\})^{(V)}$. Then $U \in V_x$ and is open by the previous lemmas. Let $U' = U \cup \{x\}$ (so $U = U' - \{x\}$, since $x \notin U$). Since $U' \in V_x$ and $U' \in V_y$ for all $y \in U$, U' is also open. So, x has an open neighborhood U' such that $U' - \{x\}$ is also open. Therefore, $\mathrm{Tp}\, V$ is locally T_1.

Now let $V' = \mathrm{It}\,\mathrm{Tp}\, V$. We will show $V = V'$. Take any $x \in X$ and suppose $C \in V_x$. Then $C - \{x\} \in V_x$ since V is irreflexive, and, letting $C' = (C - \{x\})^{(V)}$, we have that C' is open and $C' \in V_x$ by the previous lemmas. We also have that $C' \cup \{x\}$ is open (that is, $C' \cup \{x\} \in \mathrm{Tp}\, V$), so $C' = (C' \cup \{x\}) - \{x\} \in V'_x$; from $C' \subseteq C$, we have $C \in V'_x$. For the other direction, suppose $D \in V'_x$. Then there is some $U \in \mathrm{Tp}\, V$ such that $x \in U$ and $U - \{x\} \subseteq D$. Now, $U \in V_x$ by the definition of $\mathrm{Tp}\, V$, so $U - \{x\} \in V_x$ since V is irreflexive, so $D \in V_x$ since $U - \{x\} \subseteq D$. Therefore $V = V'$. □

7.7 Anonymity in **R**

To adapt Galvin's setting to one-way visibility on **R**, define $t^x : \mathbf{R} \to \mathbf{R}$ for $x \in \mathbf{R}$ by $t^x(a) = a + x$. A predictor P for a hat problem with $A = \mathbf{R}$ is anonymous if $P(f \circ t^x) = P(f) \circ t^x$ for every $x \in \mathbf{R}$ and coloring f.

It is a straightforward exercise to verify that the proof of Theorem 6.1.1 can be adapted to prove that there is an anonymous predictor for full one-way visibility on **R** that guarantees the set of errors is well ordered—that is, the predictor in Theorem 7.3.1 can additionally be taken to be anonymous. (One has to be wary of the fact that periodicity is a more subtle concept when the domain is **R**, but this turns out not to be a problem.) One might speculate that in Theorem 7.3.2 as well, one can take the predictor to be anonymous. However, this is not the case. In fact, we can construct a single coloring that makes all anonymous predictors fail.

Theorem 7.7.1. *For the hat problem of Theorem 7.3.2 (where agents see only the "recent past"), there exists a coloring f under which, for any anonymous predictor, the set of incorrect guesses is dense.*

Proof. Let $D = \{ k/2^m : k \in \mathbf{Z},\ m \in \omega \}$ denote the set of dyadic rationals. Define the function $\lambda : D \to \omega$ by $\lambda(j/2^m) = m$ (where $j/2^m$ is in lowest terms), and define the coloring $f : \mathbf{R} \to 2$ by $f(x) = \lambda(x) \bmod 2$ for $x \in D$, and $f(x) = 0$ for $x \notin D$. For $i = 0, 1$, let $D_i = D \cap f^{-1}(i)$, and note that both D_0 and D_1 are dense. We will show that for any anonymous predictor, all agents in D must guess the same under f; since both D_0 and D_1 are dense, it will follow that there must be a dense set of incorrect guesses.

For $x, y \in D$, $\lambda(x + y) \leq \max(\lambda(x), \lambda(y))$, and $\lambda(-x) = \lambda(x)$; it follows that for $x, y \in D$, if $\lambda(x) < \lambda(y)$, then $\lambda(x + y) = \lambda(y)$: $\lambda(x + y) \leq \lambda(y)$ is immediate, and $\lambda(x + y) < \lambda(y)$ would yield $\lambda(y) = \lambda(x + y + (-x)) \leq \max(\lambda(x + y), \lambda(x)) < \lambda(y)$, a contradiction.

Take any $x \in D$. We claim that there exists $\varepsilon > 0$ such that $(f \circ t^x)|(-\varepsilon, 0) = f|(-\varepsilon, 0)$. Let $m = \lambda(x)$, and let $\varepsilon = 2^{-m}$. Note that for any $y \in (-\varepsilon, 0)$, either $y \notin D$ or $\lambda(y) > \lambda(x)$. In particular, for $y \in (-\varepsilon, 0)$, $(f \circ t^x)(y) = f(y + x) = f(y)$. This establishes the claim.

Let P be any anonymous predictor. By the above claim, for any $x \in D$, we must have $P(f)(x) = P(f)(0)$. Since all agents in D make the same guess, but agents in D_0 and D_1 are assigned different colors, it must be the case that all agents in D_0 or all agents in D_1 guess incorrectly, so the set of incorrect guesses is dense. □

7.8 Open Questions

The following may be an easier version of Question 3.6.1.

Question 7.8.1. Does Corollary 7.2.4, quantified over all X and Y, imply AC over ZF?

Question 7.8.2. In the "recent past" case of Theorems 7.3.2 and 7.7.1, how well can an anonymous predictor do? Is it possible to guarantee that the set of errors is countable? Meager? Measure zero?

Given a family $T \subseteq {}^A A$, say that a predictor is *T-anonymous* if $P(f \circ t) = P(f) \circ t$ for every $t \in T$ and coloring f.

Question 7.8.3. What can be said about T-anonymous predictors? In particular, letting T be the set of order-preserving bijections on $\mathbf{R}$ (equivalently, the automorphism group of the $\mathbf{R}$ under the downward topology), how well can T-anonymous predictors do in the context of Theorem 7.3.1 (where visibility is given by the downward topology on $\mathbf{R}$)?

Chapter 8
Universality of the μ-Predictor

8.1 Background

Assume X is a topological space and Y is an arbitrary set. Then for every well-ordering $\preceq$ of $^X Y$, we let $M_\preceq$ denote the μ-predictor $\langle \mu_x : x \in X \rangle$. That is, $\mu_x(f) = \langle f \rangle_x(x)$ where $\langle f \rangle_x$ is the $\preceq$-least element of the equivalence class $[f]_x$ of f with respect to $\approx_x$, and $f \approx_x g$ iff f and g agree on a deleted neighborhood of x.

Theorem 7.2.6 showed that if X is a T_0 space, then for every well-ordering $\preceq$ of $^X Y$, the μ-predictor $M_\preceq$ is a scattered-error predictor. Here, we show that *every* scattered-error predictor in a T_0 space is an instance of the μ-predictor. This chapter is largely taken from [H13].

Theorem 8.1.1. *Suppose X is T_0 and P is a scattered-error predictor for $^X Y$. Then $P = M_\preceq$ for some well-ordering $\preceq$ of $^X Y$.*

We actually prove something slightly stronger that involves a more generalized kind of visibility to allow finer control over the information available to predictors; this generality is used in Sects. 8.5 and 8.6.

Definition 8.1.2. A *notion of visibility* $\equiv$ assigns to each $x \in X$ an equivalence relation $\equiv_x$ on $^X Y$. A function $P : {}^X Y \to {}^X Y$ *respects* $\equiv$ if, for every $x \in X$, $P(f)(x) = P(g)(x)$ whenever $f \equiv_x g$ In this case, we call P a *predictor under* $\equiv$.

Most notably, the relations $\approx_x$ above give a notion of visibility $\approx$, which we take as our default if no other notion is specified.

In any notation defined in terms of $\approx$, we add a superscript to indicate the use of $\equiv$ in place of $\approx$; for example, $[f]_x^\equiv$ is the equivalence class of f under $\equiv_x$, and $M_\preceq^\equiv f(x) = \langle f \rangle_x^\equiv(x)$. Naturally, we say that $\equiv$ *refines (coarsens)* $\equiv'$ if $\equiv_x$ refines (coarsens) $\equiv'_x$ for each $x \in X$. Note that if $\equiv$ refines $\equiv'$, then any predictor under $\equiv'$ is also a predictor under $\equiv$. We can now state the stronger version of Theorem 8.1.1.

C.S. Hardin and A.D. Taylor, *The Mathematics of Coordinated Inference*,
Developments in Mathematics 33, DOI 10.1007/978-3-319-01333-6_8,
© Springer International Publishing Switzerland 2013

Theorem 8.1.3. *Suppose X is T_0 and P is a scattered-error predictor for $^X Y$. Then there exists a well-ordering $\preceq$ of $^X Y$ such that for any notion of visibility $\equiv$ that coarsens $\approx$ and which P respects, $P = M_{\preceq}^{\equiv}$.*

In addition to characterizing the scattered-error predictors for T_0 spaces, the above results suggest a certain naturality of the μ-predictor. They also give some progress toward determining the strength of Corollary 7.2.4; in particular, does it imply AC over ZF? Our proofs of Theorems 8.1.1 and 8.1.3 are carried out in ZFC, but we examine what *can* be done in ZF in Sect. 8.7.

To prove these results, we first need to take a closer look at scattered sets.

8.2   Scattered Sets

Definition 8.2.1. For $A \subseteq X$, let

$$\lim A = \{\, x \in X : \text{every deleted neighborhood of } x \text{ intersects } A \,\}.$$

Define $A^{\bullet} = A \cap \lim A$, and define $A^{(\alpha)}$ for ordinals α by

$$A^{(0)} = A$$
$$A^{(\alpha+1)} = \left(A^{(\alpha)}\right)^{\bullet}$$
$$A^{(\lambda)} = \bigcap_{\alpha<\lambda} A^{(\alpha)} \quad (\lambda \text{ a limit ordinal}).$$

The *rank* of A is the least ordinal $\rho(A)$ such that $A^{(\rho(A)+1)} = A^{(\rho(A))}$; we call $A^{(\rho(A))}$ the *kernel* of A.

This is very similar to Cantor-Bendixson derivatives and rank, except that we have $A^{\bullet} = A \cap \lim A$, while the Cantor-Bendixson derivative of A is $\lim A$.

Proposition 8.2.2. *A set is scattered iff its kernel is $\emptyset$.*

Proposition 8.2.3. *In the downward (upward) topology on a partial order, the scattered sets are the well-founded (co-well-founded) sets.*

Proposition 8.2.4. *If sets $U_i \subseteq X$, $i \in I$, are open and $\Sigma \subseteq \cup_i U_i$, then Σ is (weakly) scattered iff $\Sigma \cap U_i$ is (weakly) scattered for each $i \in I$.*

Proposition 8.2.5. *The family $\mathcal{I}$ of weakly scattered subsets of X forms an ideal; i.e., $\emptyset \in \mathcal{I}$, $(A \in \mathcal{I} \wedge B \subseteq A) \Rightarrow B \in \mathcal{I}$, and $A, B \in \mathcal{I} \Rightarrow A \cup B \in \mathcal{I}$.*

(In T_0 spaces, this ideal is the same as the family of scattered sets. In non-T_0 spaces, the latter is not an ideal: If x_0 and x_1 witness that X is not T_0, then $\{x_0\}$ and $\{x_1\}$ are scattered but $\{x_0, x_1\}$ is not.)

We define the relation $=^\dagger$ on $^X Y$ by $f =^\dagger g$ iff $f \Delta g$ is weakly scattered.

In light of the above proposition, this is an equivalence relation, and we use $[f]^\dagger$ to denote the equivalence class of f under $=^\dagger$.

If X is a topological space and $x \in X$, we let $\bar{x}$ denote the set of all $y \in X$ such that every neighborhood of x contains y.

Lemma 8.2.6. *If X is T_0, $\Sigma \subseteq X$ is scattered and $x \in X$, then x has a neighborhood V such that $V \cap \Sigma \cap \bar{x} - \{x\} = \emptyset$.*

Proof. Let $\Sigma' = \Sigma \cap \bar{x} - \{x\}$. If $\Sigma' = \emptyset$, then we can let V be any neighborhood of x. Otherwise, there exists $y \in \Sigma'$ with neighborhood W such that $W \cap \Sigma' = \{y\}$. Since $x \leq y$, every neighborhood of y contains x; in particular, W is a neighborhood of x, and since X is T_0 and $x \neq y$, x has a neighborhood U such that $y \notin U$. Let $V = W \cap U$. Then V is a neighborhood of x disjoint from Σ', so $V \cap \Sigma \cap \bar{x} - \{x\} = \emptyset$.   □

8.3 Dynamics of Scattered-Error Predictors

Throughout this section, we assume X is T_0.

Proposition 8.3.1. *If U is open and $f|U = f'|U$, then $P(f)|U = P(f')|U$ for any predictor P.*

Lemma 8.3.2. *Suppose $x \in X$ and $f, f' \in {}^X Y$ such that $f \Delta f' \subseteq \{x\}$. Then for any scattered-error predictor P, x has a neighborhood V such that $(P(f) \Delta P(f')) \cap V = \emptyset$.*

Proof. Let $\Sigma = P(f) \Delta P(f')$, which is scattered since $P(f) =^\dagger f =^\dagger f' =^\dagger P(f')$. By Lemma 8.2.6, let V be a neighborhood of x such that $V \cap \Sigma \cap \bar{x} - \{x\} = \emptyset$. We claim that $\Sigma \cap V = \emptyset$; for this, it suffices to show that $\Sigma \subseteq \bar{x} - \{x\}$. If $y \notin \bar{x}$, then y has a neighborhood U with $x \notin U$; U witnesses $f \approx_y f'$, so $P(f)(y) = P(f')(y)$ and hence $y \notin \Sigma$. Also, $f \approx_x f'$, so $x \notin \Sigma$. This establishes the claim and we now have $(P(f) \Delta P(f')) \cap V = \Sigma \cap V = \emptyset$.   □

Lemma 8.3.3. *If P is a scattered-error predictor, then every equivalence class of $=^\dagger$ contains exactly one fixed point of P.*

Proof. Let $h \in {}^X Y$. We first show that P has at most one fixed point in $[h]^\dagger$. Suppose $f, f' \in [h]^\dagger$ are distinct fixed points. Then, $f \Delta f'$ is nonempty but scattered. Let $x \in f \Delta f'$ with neighborhood V such that $(f \Delta f') \cap V = \{x\}$. Then $f \approx_x f'$, so $f(x) = P(f)(x) = P(f')(x) = f'(x)$, a contradiction. So P has at most one fixed point in $[h]^\dagger$. It remains to be shown that a fixed point of P in $[h]^\dagger$ exists.

Let

$$\mathcal{A} = \{ U \subseteq X : U \text{ is open and } \exists f \in [h]^\dagger \ P(f)|U = f|U \}.$$

We first show that no proper subset of X is a maximal element of $\mathcal{A}$, and then show that $\mathcal{A}$ is closed under arbitrary unions.

Take any $U \in \mathcal{A}$, and let $f \in [h]^\dagger$ be such that $P(f)|U = f|U$. Assume $P(f) \Delta f \neq \emptyset$ (otherwise, f is a fixed point and we are done). Since $P(f) \Delta f$ is scattered, there exists $x \in P(f) \Delta f$ with neighborhood V such that $(P(f) \Delta f) \cap V = \{x\}$. Let $f' = f[P(f)(x)/x]$, where $f' = f[b/a]$ is defined by $f'(x) = b$ if $x = a$, and $f'(x) = f(x)$ otherwise. By Lemma 8.3.2, shrinking V if necessary, we can assume without loss of generality that $(P(f') \Delta P(f)) \cap V = \emptyset$. For $z \in V - \{x\}$, we have $P(f')(z) = P(f)(z) = f(z) = f'(z)$, and we have $P(f')(x) = P(f)(x) = f'(x)$; this yields $P(f')|V = f'|V$.

For $z \in U$, Proposition 8.3.1 gives us $P(f')(z) = P(f)(z) = f(z) = f'(z)$, so $P(f')|U = f'|U$. We now have $P(f')|(U \cup V) = f'|(U \cup V)$, so f' witnesses that $(U \cup V) \in \mathcal{A}$. Note that U is a proper subset of $U \cup V$, since $x \in V - U$. So, no proper subset of X is maximal in $\mathcal{A}$.

We now show that $\mathcal{A}$ is closed under arbitrary unions. Suppose we have sets $U_i \in \mathcal{A}$ with associated functions $f_i \in [h]^\dagger$, for i in some index set I. We first show that the partial functions $f_i|U_i$ are compatible. If not, then there are $j, k \in I$ such that, letting $V = U_j \cap U_k$, the set $\Sigma = (f_j \Delta f_k) \cap V$ is nonempty. Since Σ is scattered, there exists $x \in \Sigma$ with neighborhood V such that $\Sigma \cap V = \{x\}$. Then $f_j \approx_x f_k$, so $Sf_j(x) = Sf_k(x)$, so $f_j(x) = Sf_j(x) = Sf_k(x) = f_k(x)$, a contradiction.

Let $U = \bigcup_{i \in I} U_i$. Since the partial functions $f_i|U_i$ are compatible, their union is a function $f : U \to Y$. Extend f to a function $f : X \to Y$ by letting $f|(X - U) = h|(X - U)$. Take any $x \in U$, and let $i \in I$ be such that $x \in U_i$; noting that $f \approx_x f_i$, we have $P(f)(x) = P(f_i)(x) = f_i(x) = f(x)$. It follows that $P(f)|U = f|U$, so f witnesses that $U \in \mathcal{A}$, provided $f =^\dagger h$, which follows from Proposition 8.2.4. This establishes the claim that $\mathcal{A}$ is closed under arbitrary unions.

Since $\mathcal{A}$ is closed under arbitrary unions, and no proper subset of X is maximal in $\mathcal{A}$, it follows that $X \in \mathcal{A}$. The f witnessing $X \in \mathcal{A}$ is a fixed point of P.   □

Lemma 8.3.4. *Given any $f : X \to Y$ and $D \subseteq X$, there exists $f' : X \to Y$ such that $f'|D = f|D$ and $P(f') \Delta f' \subseteq D$.*

The idea in the following proof is that fixing $f|D$ induces a scattered-error predictor for $^{(X-D)}Y$ to which we can apply the above lemma.

Proof. Let $X_0 = X - D$ with the subspace topology. For any function $h : X_0 \to Y$, define $\hat{h} : X \to Y$ by $\hat{h}|X_0 = h$, $\hat{h}|D = f|D$. Define the predictor P_0 for ^{X_0}Y by $P_0(g) = P(\hat{g})|X_0$. It follows from the fact that P is a scattered-error predictor (for ^XY) that P_0 is a scattered-error predictor (for ^{X_0}Y), so by the previous lemma there is a function $h : X_0 \to Y$ such that $h =^\dagger f|X_0$ and $P_0(h) = h$. Then $P(\hat{h})|X_0 = P_0(h) = h = \hat{h}|X_0$, so $P(\hat{h}) \Delta \hat{h} \subseteq D$, and $\hat{h}|D = f|D$.   □

8.4 Getting an Ordering from a Predictor

Throughout this section, we assume X is T_0.

Fix a scattered-error predictor P, and let $\preceq$ be any well-ordering of ^XY such that $\rho(P(f)\Delta f) < \rho(P(f')\Delta f') \Rightarrow f \prec f'$. We will show that the resulting $M_\preceq$ coincides with P (and, more generally, $M_\preceq^\equiv = P$ for appropriate $\equiv$). To get some intuition for why this will work, if we have $P(\langle f\rangle_x)(x) = \langle f\rangle_x(x)$, then it will follow that $P(f)(x) = P(\langle f\rangle_x)(x) = \langle f\rangle_x(x) = M_\prec(f)(x)$, which we want. In order to favor functions g where $P(g)$ and g agree at x, but without making specific reference to x (since we have one ordering $\preceq$ that is used at all points), we simply favor functions g where $P(g)$ and g agree often. One way to say that $P(g)$ and g agree often is to say that $\rho(P(g)\Delta g)$ is small. By placing functions g with small values of $\rho(P(g)\Delta g)$ early in the ordering, we will tend to get $P(\langle f\rangle_x)(x) = \langle f\rangle_x(x)$. That this is not just a tendency, but *always* happens, is worked out in the details that follow.

Suppose $\equiv$ is a notion of visibility that coarsens $\approx$ but which is still respected by P in the sense that $f \equiv_x g \Rightarrow P(f)(x) = P(g)(x)$.

Lemma 8.4.1. *Take any $f \in {}^XY$ and $x \in X$, and let $g = \langle f\rangle_x^\equiv$. Then for any neighborhood V of x, $\rho((P(g)\Delta g) \cap V - \{x\}) = \rho(P(g)\Delta g)$.*

Proof. It is immediate that $\rho((P(g)\Delta g) \cap V - \{x\}) \leq \rho(P(g)\Delta g)$. Suppose for a contradiction that $\rho((P(g)\Delta g) \cap V - \{x\}) < \rho(P(g)\Delta g)$. Let $g' = g[P(g)(x)/x]$. By Lemma 8.3.2, let V' be a neighborhood of x such that

$$(P(g)\Delta P(g')) \cap V' = \emptyset. \tag{8.1}$$

Without loss of generality, $V' \subseteq V$. By Lemma 8.3.4, there is a $g'' \in {}^XY$ such that $g'|V' = g''|V'$ and $P(g'')\Delta g'' \subseteq V'$. Note that $P(g)|V' = P(g')|V' = P(g'')|V'$ by (8.1) and Proposition 8.3.1.

We claim that $P(g'')\Delta g'' \subseteq (P(g)\Delta g)\cap V-\{x\}$. Take any $z \in P(g'')\Delta g''$. Then $z \in V' \subseteq V$, since $P(g'')\Delta g'' \subseteq V'$. Note that $g \approx_x g' \approx_x g''$ (the former because $g\Delta g' = \{x\}$, the latter because $g'|V' = g''|V'$), so $P(g'')(x) = P(g)(x) = g'(x)$, so $x \notin P(g'')\Delta g''$, so $z \neq x$. Also, $P(g)(z) = P(g'')(z) \neq g''(z) = g'(z) = g(z)$, so $z \in P(g)\Delta g$. We now have $z \in (P(g)\Delta g) \cap V - \{x\}$. This establishes the claim.

It follows that $\rho(P(g'')\Delta g'') \leq \rho((P(g)\Delta g) \cap V - \{x\}) < \rho(P(g)\Delta g)$, so $g'' \prec g$. Note, however, that $g'' \approx_x g \equiv_x f$, so $g'' \equiv_x f$, so g is not the $\preceq$-least element of $[f]_x^\equiv$, a contradiction. □

Lemma 8.4.2. *Let Σ be a scattered set and suppose $x \in X$ is such that $\rho(\Sigma \cap V - \{x\}) = \rho(\Sigma)$ for every neighborhood V of x. Then $x \notin \Sigma$.*

Proof. Let $\sigma = \rho(\Sigma)$. Let γ be minimal such that $x \notin \Sigma^{(\gamma)}$. Note that $\gamma \leq \sigma$ since $\Sigma^{(\sigma)} = \emptyset$. Note also that γ cannot be a limit ordinal (since, for any limit ordinal λ, any point absent from $\Sigma^{(\lambda)}$ is already absent from $\Sigma^{(\alpha)}$ for some $\alpha < \lambda$).

Suppose for a contradiction $x \in \Sigma$. Then $\gamma \neq 0$, so $\gamma = \beta + 1$ for some β. Then, $x \in \Sigma^{(\beta)}$ and x has a neighborhood V such that $\Sigma^{(\beta)} \cap V - \{x\} = \emptyset$. Then $(\Sigma \cap V - \{x\})^{(\beta)} = \emptyset$, so $\rho(\Sigma \cap V - \{x\}) \leq \beta < \sigma = \rho(\Sigma \cap V - \{x\})$, a contradiction. Therefore, $x \notin \Sigma$. $\square$

Lemma 8.4.3. *Take any $f \in {}^X Y$ and $x \in X$, and let $g = \langle f \rangle_x^{\equiv}$. Then $P(g)(x) = g(x)$.*

Proof. Let $\Sigma = P(g) \Delta g$, a scattered set. By Lemma 8.4.1, for any neighborhood V of x, $\rho(\Sigma \cap V - \{x\}) = \rho(\Sigma)$. By Lemma 8.4.2, $x \notin \Sigma$, so $P(g)(x) = g(x)$. $\square$

Proof of Theorem 8.1.3. With P, $\preceq$, and $\equiv$ as above, take any $f \in {}^X Y$ and $x \in X$. Let $g = \langle f \rangle_x^{\equiv}$. By the previous lemma, $P(g)(x) = g(x)$. Since $g \equiv_x f$, we have $P(g)(x) = P(f)(x)$. Then $M_{\preceq}^{\equiv} f(x) = g(x) = P(g)(x) = P(f)(x)$. Therefore, $P = M_{\preceq}^{\equiv}$. $\square$

Theorem 8.1.1 follows as the special case where $\equiv$ is $\approx$.

8.5   Visibility Graphs

So far, the results of this chapter have applied to the topological context. We now consider how the results carry over when the notion of visibility is given by a visibility graph. Given a visibility graph V on X, let $\sim$ be the resulting notion of visibility (that is, $f \sim_x g$ iff $f(y) = g(y)$ for all $y \in V(x)$). Observe that if V is transitive and we put the upward topology on X derived from the partial ordering where $x < y$ iff xVy, then $\sim$ and $\approx$ coincide (where $\approx$ is as in Definition 7.2.1). In short, transitive visibility is a special case of the topological context. Recalling that the scattered sets in the upward topology on a partial order are the co-well-founded sets, the role played by scattered sets in previous sections is played by co-well-founded sets below.

What we are able to show is that when V is acyclic, "good" predictors (when they exist at all) are all special cases of the μ-predictor. In the following, V^+ denotes the transitive closure of V.

Theorem 8.5.1. *Let V be an acyclic visibility graph on X, and suppose that P is a predictor for V such that $P(f) \Delta f$ is co-well-founded in V^+ for all $f \in {}^X Y$. Then $P = M_{\preceq}$ for some well-ordering $\preceq$ of ${}^X Y$. (Now, of course, $M_{\preceq}$ refers to $M_{\preceq}^{\sim}$, not $M_{\preceq}^{\approx}$.)*

Proof. As V is acyclic, V^+ is a strict partial order of X. Consider X as a topological space under the upward topology induced by V^+, and let $\approx$ be the resulting notion of visibility. Note that $\approx$ refines $\sim$, so P respects $\approx$. Also, as noted above, the scattered sets coincide with the sets co-well-founded in V^+. So, we can consider P as a scattered-error predictor under $\approx$. Applying Theorem 8.1.3, let $\preceq$ be a well-ordering of ^XY such that $P = M_{\preceq}^{\equiv}$ for any $\equiv$ that coarsens $\approx$ and which P respects. In particular, this applies when $\equiv$ is $\sim$, so $P = M_{\preceq}^{\sim}$. $\qquad\square$

A case of particular interest is finite-error predictors. The question of which relations V admit a finite-error predictor is an ongoing one; specifically, we would like to know whether or not the following are equivalent for $|Y| \geq 2$:

(i) V admits a finite-error predictor;
(ii) There is no sequence of distinct $x_0, x_1, \ldots \in X$ such that $x_i V x_j$ fails for every $i \leq j$.

The direction (i)$\Rightarrow$(ii) always holds, and (ii)$\Rightarrow$(i) is known to hold when X is countable (Theorem 4.3.1) or V is transitive (Theorem 5.3.1). Also, if (ii)$\Rightarrow$(i) holds for acyclic V, then it holds for all V (since intersecting V with a well-ordering of X makes V acyclic while preserving (ii)). The following corollary tells us that, in the acyclic case, we can restrict our attention to instances of the μ-predictor when seeking a finite-error predictor.

Corollary 8.5.2. *Suppose V is an acyclic visibility graph on X and that P is a finite-error predictor for V. Then $P = M_{\preceq}$ for some well-ordering $\preceq$ of ^XY.*

Proof. Finite sets are necessarily co-well-founded in any partial order, so Theorem 8.5.1 applies. $\qquad\square$

There is no hope of extending this to visibility graphs that contain cycles (except in degenerate cases where there are no such S to begin with, or $|Y| \leq 1$), as the following proposition shows. Say that two predictors P and P' are *almost the same* if $P(f)\Delta P'(f)$ is finite for all $f \in {}^XY$. Note that if P and P' are almost the same and I is a non-principal ideal, then P is an I^*-predictor iff P' is an I^*-predictor.

Theorem 8.5.3. *Suppose V has a cycle, P is a predictor for V, and $|Y| \geq 2$. Then there exists a predictor P' that is almost the same as P and is not a special case of the μ-predictor.*

Proof. We will use the cycle to construct P' in a way that guarantees at least one error. Such a predictor cannot be a special case of the μ-predictor, because there is always at least one function that makes the μ-predictor correct everywhere: for any well-ordering $\preceq$ of ^XY, if f_0 is the first function in the ordering, $M_{\preceq}(f_0) = f_0$.

Let $x_0 V x_1 V \cdots V x_{k-1} V x_0$ be a cycle of V. Let $d : Y \to Y$ be such that $d(y) \neq y$ for all $y \in Y$. For $f \in {}^XY$, we define

$$P'(f)(x) = \begin{cases} f(x_{i+1}) & \text{if } x = x_i,\, i < k-1, \\ d(f(x_0)) & \text{if } x = x_{k-1}, \\ P(f)(x) & \text{otherwise.} \end{cases}$$

Informally, in P', all agents in the cycle other than x_{k-1} assume their hat color is the same as the color of the next agent in the cycle, while x_{k-1} assumes it is not;

everywhere else, P' agrees with P (so P' is almost the same as P). This guarantees at least one error: if P' were correct at every point in the cycle, we would have $f(x_0) = f(x_1) = \cdots = f(x_{k-1}) = d(f(x_0)) \neq f(x_0)$, a contradiction.   □

8.6   Variations on the μ-Predictor

In this section we consider modified versions of the μ-predictor such as the μ^*-predictor (which ignores finite differences) from Chap. 7. One virtue of the μ^*-predictor is that while the proof of Theorem 7.2.6 is about one page, the proof of the analogous result for the μ^*-predictor, Theorem 7.2.3, is 11 lines; that gives the μ^*-predictor, perhaps, a greater claim to being the "right" approach. Another virtue of the μ^*-predictor is that its willingness to overlook certain minor differences makes it work in some contexts where the μ-predictor can fail. For example, if one lets V be the complement of the identity relation on a set X, then the μ^*-predictor will always be finite-error, but the μ-predictor will typically not be; also, as noted below, the μ^*-predictor is weakly scattered–error even in non-T_0 spaces.

Taking this idea further, we can consider the $\mu^\dagger$-predictor, which ignores weakly scattered sets of differences. (This only makes sense in the topological context. Though we can make sense of the μ^*-predictor when working with visibility graphs, we only consider the topological case below.)

Formally, under a given notion of visibility $\approx$, let $\approx^*$ be the finest coarsening of $\approx$ in which each $\approx_x^*$ respects $=^*$; define $\approx^\dagger$ similarly. For a given well-ordering $\preceq$ of ^XY, the μ^*-predictor under $\approx$ is $M_{\preceq}^{\approx^*}$, while the $\mu^\dagger$-predictor under $\approx$ is $M_{\preceq}^{\approx^\dagger}$.

Given a scattered-error predictor P that respects $=^*$ (or $=^\dagger$), we already know (provided X is T_0) that P must be a special case of the μ-predictor. By Theorem 8.1.3, we can also say that P must be a special case of the μ^*-predictor (or, respectively, the $\mu^\dagger$-predictor).

Much else of what we already know about the μ-predictor also carries immediately over to the μ^*- and $\mu^\dagger$-predictors. As detailed below, the μ^*- and $\mu^\dagger$-predictors can be obtained as special cases of the μ-predictor under a finer topology. Our only concern is that, when we refine the topology, we might introduce new weakly scattered sets, so that while the μ^*- or $\mu^\dagger$-predictor is weakly scattered–error with respect to the finer topology, perhaps it is not weakly scattered–error with respect to the original topology. We show below that, for the refinements under consideration, no new weakly scattered sets are introduced.

Definition 8.6.1. Given a topology $\mathcal{T}$ on X, let $S(\mathcal{T})$ denote the ideal of sets that are weakly scattered with respect to $\mathcal{T}$, let $\mathcal{T}^*$ be the coarsest refinement of $\mathcal{T}$ containing all cofinite sets (equivalently, the coarsest T_1 refinement of $\mathcal{T}$), and let $\mathcal{T}^\dagger$ be the coarsest refinement of $\mathcal{T}$ containing the complements of sets in $S(\mathcal{T})$.

Observe that the μ^*-predictor, under $\mathcal{T}$, is realized as the μ-predictor under $\mathcal{T}^*$; likewise for the $\mu^\dagger$-predictor and $\mathcal{T}^\dagger$. Note that $\mathcal{T}^*$ and $\mathcal{T}^\dagger$ are always T_1, even if

$\mathcal{T}$ is not T_0 (once the theorem below is proved, this shows that the μ^*-predictor and the $\mu^\dagger$-predictor are weakly scattered–error in any space).

Proposition 8.6.2. $\mathcal{T} \subseteq \mathcal{T}^* \subseteq \mathcal{T}^\dagger = \{U - K : U \in \mathcal{T} \wedge K \in S(\mathcal{T})\}$.

Theorem 8.6.3. $S(\mathcal{T}) = S(\mathcal{T}^*) = S(\mathcal{T}^\dagger)$.

Proof. By $\mathcal{T} \subseteq \mathcal{T}^* \subseteq \mathcal{T}^\dagger$, the inclusions $S(\mathcal{T}) \subseteq S(\mathcal{T}^*) \subseteq S(\mathcal{T}^\dagger)$ are trivial, so we must show $S(\mathcal{T}^\dagger) \subseteq S(\mathcal{T})$. Suppose $\Sigma \in S(\mathcal{T}^\dagger)$, and take any nonempty $\Sigma' \subseteq \Sigma$. We must show that Σ' has a point that is weakly isolated with respect to $\mathcal{T}$. Let $x \in \Sigma'$ with neighborhood $V \in \mathcal{T}^\dagger$ such that $\Sigma' \cap V$ is finite. Then $V = U - K$ for some $U \in \mathcal{T}$ and $K \in S(\mathcal{T})$. Without loss of generality, $K \subseteq U$ (so $U = V \cup K$). If $\Sigma' \cap U$ is finite, we are done. Otherwise, $\Sigma' \cap K$ must be infinite; in particular, it is a nonempty subset of $K \in S(\mathcal{T})$, so there exists some $y \in \Sigma' \cap K$ with neighborhood $W \in \mathcal{T}$ such that $W \cap \Sigma' \cap K$ is finite. One can now verify that $U \cap W \in \mathcal{T}$ is a neighborhood of $y \in \Sigma'$ that weakly isolates y from Σ'. Therefore, $\Sigma \in S(\mathcal{T})$. □

8.7 Results Without the Axiom of Choice

We would like to know whether Corollary 7.2.4 (quantified over all X and Y) implies AC over ZF. For this purpose, the main results are not immediately of any use, since they are theorems of ZFC. Though all of Sect. 8.2 can be carried out in ZF, we appeal to AC at the beginning of Sect. 8.4 when extending $\preceq$ to a well-ordering. What happens if we skip that step?

Suppose that, at the beginning of Sect. 8.2, we had chosen to let $f \prec f' \Leftrightarrow \rho(P(f) \triangle f) < \rho(P(f') \triangle f')$, without extending to a well-ordering. This would be a well-founded partial order of ${}^X Y$; it would not be total (except in degenerate cases), but it would be total enough (when X is T_0, at least) to uniquely determine $M_{\preceq}^{\equiv}$: roughly speaking, if it did not uniquely determine $M_{\preceq}^{\equiv}$, then our proof of Theorem 8.1.1 would not work, since it uses an arbitrary extension of the above ordering to $\preceq$. A more rigorous justification follows.

Rather than letting $\langle f \rangle_x^{\equiv}$ be the $\preceq$-least element of $[f]_x^{\equiv}$, we now define $\langle f \rangle_x^{\equiv}$ to be the set of $\preceq$-minimal elements of $[f]_x^{\equiv}$. Fix some $y_0 \in Y$ (the case $Y = \emptyset$ is uninteresting). We define $M = M_{\preceq}^{\equiv}$ as follows: if every $g \in \langle f \rangle_x^{\equiv}$ agrees on the value of $g(x)$, we take this to be $M(f)(x)$; otherwise, we let $M(f)(x) = y_0$. (This latter case will never occur, but we cannot assume that yet.) In the statements of Lemmas 8.4.1 and 8.4.3 and the proof of Theorem 8.1.3, $g = \langle f \rangle_x^{\equiv}$ becomes $g \in \langle f \rangle_x^{\equiv}$. At the end of the proof of Lemma 8.4.1, the contradiction is now that g is not $\preceq$-minimal in $[f]_x^{\equiv}$, rather than "g is not the $\preceq$-least element of $[f]_x^{\equiv}$." With these modifications, we still reach the conclusion $P = M_{\preceq}^{\equiv}$ in the proof of Theorem 8.1.3. (Also note that, with the modified version of Lemma 8.4.3, every $g \in \langle f \rangle_x^{\equiv}$ agrees on the value of $g(x)$: for $g, g' \in [f]_x^{\equiv}$, we have $g(x) = P(g)(x) = P(g')(x) = g'(x)$; so, the y_0 case above never occurs.)

8.8 Open Questions

We have seen that, in the context of T_0 spaces, every scattered-error predictor is an instance of the μ-predictor, and that every instance of the μ-predictor is scattered-error. Transitive visibility graphs can be seen as a special case of this. However, nontransitive visibility graphs are not as well understood. What we have shown is that, for an acyclic visibility graph V, every *good* predictor (that is, one guaranteeing that the set of errors is co-well-founded in V^+) is an instance of the μ-predictor; it is not always the case, though, that every well-ordering $\preceq$ makes $M_\preceq$ a good predictor. First, some relations admit no good predictor at all (for example, with V the successor relation on $X = \omega$ and $|Y| \geq 2$, no predictor can guarantee even a single correct guess); second, even when good predictors exist, $M_\preceq$ will be good for some choices of $\preceq$, but typically not all when V is nontransitive.

So, two questions arise:

Question 8.8.1. Which visibility graphs admit good predictors?

Question 8.8.2. When a visibility graph admits at least one good predictor, which well-orderings $\preceq$ make $M_\preceq$ a good predictor?

Even if we cannot answer the latter question fully, can we at least find a way to construct $\preceq$ such that, if there is any good predictor at all, then $M_\preceq$ is good?

Currently, the only known technique for producing good predictors based on the μ-predictor for nontransitive visibility graphs is to voluntarily coarsen the notion of visibility to one that is more cooperative, without coarsening it too much. For example, given a nontransitive visibility graph V, we can often find a transitive $T \subseteq V$ that is "close" to V in some sense, and use the μ-predictor with T as our notion of visibility, as in Sect. 4.3. In [H10], an example is given for which that approach cannot be made to work; specifically, a nontransitive V is constructed that holds some promise for admitting a finite-error strategy, but for which no transitive subrelation admits a finite-error strategy. Yet we know from Corollary 8.5.2 that if any finite-error strategy exists, it can be realized as a special case of the μ-predictor. This is some of our motivation for the above questions: in situations where restricting to a transitive subrelation is not an option, we would like a way of constructing orderings $\preceq$ that make the μ-predictor perform well even in the absence of transitivity.

In the case of visibility graphs with a cycle, we saw in Theorem 8.5.3 how predictors can fail to be special cases of the μ-predictor. Nevertheless, we ask the following:

Question 8.8.3. Can we identify the circumstances under which, given a predictor P, there exists an instance of the μ-predictor that is "as good" as P? For example, is it the case that if I is an ideal and $\equiv$ is a notion of visibility that admits an I^*-predictor, then there is an instance of the μ-predictor that is an I^*-predictor?

Chapter 9
Generalizations and Galois-Tukey Connections

9.1 Background

It is possible to see the hat problem on the parity relation EO as actually being a two-agent hat problem, though we must first consider a more general type of hat problem; in doing so, we uncover a close relationship with so-called Galois-Tukey connections, which arise in the study of cardinal invariants. See [Mild97] for background. Rather than requiring agents to guess their hat colors exactly, we can have a more flexible notion of when a guess is acceptable. We might, say, have **R** as the set of hat colors, and require an agent's guess to fall within a certain radius of the actual color; or the agent might be allowed to guess any meager set of reals, with the guess considered acceptable when the actual color is in that set. More generally, we can use any binary relation between two sets to specify when guesses are acceptable. Much of our notation and terminology below follows [Bar10], but see also [Bla10].

Here, we require any binary relation R to have distinguished domain R_- and codomain R_+, so $R \subseteq R_- \times R_+$. It will often be convenient to specify such a relation with a triple (R_-, R_+, R), as in $(\mathbf{R}, \mathcal{M}, \in)$ where $\mathcal{M}$ denotes the collection of meager sets of reals. A *relational hat problem* is much like an ordinary hat problem, except that rather than having a single set of hat colors, we assign to each agent x a relation R^x; a coloring assigns to each agent x an element of R^x_-; each agent x, rather than guessing a color, guesses an element of R^x_+ (so a predictor is now a function $P : \prod_{x \in X} R^x_- \to \prod_{x \in X} R^x_+$); for a color $c \in R^x_-$ and guess $d \in R^x_+$, we consider the guess correct if $c R^x d$. While we still refer to elements of R^x_- as colors, we refer to elements of R^x_+ as *co-colors*, so each agent is assigned a color (and sees the colors of visible agents) but guesses a co-color. We refer to the relations R^x as *acceptability relations*, and we refer to them collectively as R, which we call an *acceptability notion for* X.

Definition 9.1.1. For a relation R, $R^\perp$ is the complement of the converse of R. For example, $(\mathbf{R}, \mathcal{M}, \in)^\perp = (\mathcal{M}, \mathbf{R}, \not\ni)$.

C.S. Hardin and A.D. Taylor, *The Mathematics of Coordinated Inference*,
Developments in Mathematics 33, DOI 10.1007/978-3-319-01333-6_9,
© Springer International Publishing Switzerland 2013

R is *serial* if $\forall c \in R_- \; \exists d \in R_+ \; cRd$; equivalently (under AC), there is a function $f : R_- \to R_+$ with $f \subseteq R$. We say R is *decent* (not standard terminology) if both R and $R^\perp$ are serial.

In terms of hat problems, the two conditions for R being decent tell us, respectively, that there is no color that can make every guess wrong, and no guess that is correct for every color.

An important observation to make is that the main positive and negative results for the topological case (that is, the results of Sect. 7.2) still work in the context of relational hat problems, provided we are using decent acceptability relations. For a trivial diagonalization as in Theorem 7.2.5, the proof carries over unchanged: once an agent's guess is determined, we can assign a color that makes the guess wrong. The positive results carry over easily as corollaries, since guessing one's hat color exactly is always at least as hard as coming up with an acceptable guess under a given decent relation. More precisely, we do the following. Given decent relations R^x for $x \in X$, we fix a well-ordering of $\prod_{x \in X} R^x_-$ and, for each $x \in X$, fix a function $g^x : R^x_- \to R^x_+$ such that $cR^x g^x(c)$ for all $c \in R^x_-$; the resulting variant M of the μ-predictor is defined by $M(f)(x) = g^x(\langle f \rangle_x(x))$.

With visibility specified by a T_0 space (which includes transitive visibility graphs as a special case), those results were already sharp—we can get scattered-error predictors but nothing better—so the increased generality of relational hat problems provides nothing of interest there: just as the number of colors (provided it is at least two) is not relevant in ordinary hat problems for this kind of visibility, the combinatorics of the acceptability relations (provided they are decent) is not relevant in relational hat problems. For nontransitive visibility, however, the number of colors matters, and accordingly, the combinatorics of the acceptability relations is relevant. Furthermore, with the acceptability relations themselves of significant importance, one can get rich hat problems even with very simple visibility graphs; we see in Sect. 9.3, for instance, that in the case of two agents who see each other, the existence of a minimal predictor is equivalent to the existence of a Galois-Tukey connection between two particular relations.

9.2　Galois-Tukey Connections

Our starting point in this section is the following from [V93].

Definition 9.2.1. For relations A, B, a *morphism* (or *Galois-Tukey connection*) $m :$ $A \to B$ is a pair $m = (f, g)$ where $f : A_- \to B_-$, $g : B_+ \to A_+$ such that for all $a \in A_-, b \in B_+$, if $f(a)Bb$, then $aAg(b)$. When such a morphism exists, we write $A \preceq B$ and say that B is *harder* than A (and that A is *easier* than B).

We can express this in more algebraic terms.

Proposition 9.2.2. $(f, g) : A \to B$ *iff* $g \circ B \circ f \subseteq A$.

Proof. Suppose $(f, g) : A \to B$ and $x(g \circ B \circ f)z$. Then there exists $y \in B_+$ such that $x(B \circ f)y$ and $y(g)z$ (i.e., $f(x)By$ and $g(y) = z$). Then, since $f(x)By$, we have $xAg(y)$, so xAz. Therefore, $g \circ B \circ f \subseteq A$.

Suppose $g \circ B \circ f \subseteq A$. Suppose $x \in A_-$, $y \in B_+$ such that $f(x)By$. Then $x(B \circ f)y$, so $x(g \circ B \circ f)z$ where $z = f(y)$, so $xAf(y)$. □

Proposition 9.2.3. $(f, g) : A \to B$ *iff* $(g, f) : B^\perp \to A^\perp$.

Note that if $A \subseteq A'$ are relations (with the same domain and same codomain), then $A' \preceq A$ via the morphism $(\mathrm{id}_{A_-}, \mathrm{id}_{A_+})$. So the definition of $\preceq$ might at first seem "backwards"; however, $\preceq$ is more closely related to dominating sets (see Definition 9.4.1), which get smaller as edges are added to a relation.

We now justify our use of the terms "harder" and "easier." Informally, the following theorem says that if a relational hat problem admits a "good" predictor and you replace the acceptability relations with easier relations, then the new problem also admits a good predictor.

Theorem 9.2.4. *Suppose X is a set of agents with visibility graph (or proximity scheme) V; $\mathcal{I} \subseteq \mathcal{P}(X)$ is closed under formation of subsets; Q and R are acceptability notions for X; for all $x \in X$, $Q^x \preceq R^x$, witnessed by morphism (f^x, g^x); and $P : \prod_{x \in X} R^x_- \to \prod_{x \in X} R^x_+$ is a predictor (under R) guaranteeing that the set of wrong guesses is in $\mathcal{I}$. Then there exists a predictor P' : $\prod_{x \in X} Q^x_- \to \prod_{x \in X} Q^x_+$ (under Q) guaranteeing that the set of wrong guesses is in $\mathcal{I}$.*

Though the notation gets a bit messy, the basic idea behind the following proof is simple: we produce P' by placing P inside a wrapper that uses the f^x to transform what the agents see, and the g^x to transform what they guess in response.

Proof. Define $f : \prod_{x \in X} Q^x_- \to \prod_{x \in X} R^x_-$ by $f(h)(x) = f^x(h(x))$ (i.e., f takes a coloring under Q and applies the functions f^x in the natural way to yield a coloring under R). Similarly, define $g : \prod_{x \in X} R^x_+ \to \prod_{x \in X} Q^x_+$ by $g(h)(x) = g^x(h(x))$. Let $P' = g \circ P \circ f$.

Given $x \in X$ and $h, h' \in \prod_{x \in X} Q^x_-$ such that $h \approx_x h'$, it easily follows that $f(h) \approx_h f(h')$, since f acts coordinatewise, so $P(f(h))(x) = P(f(h'))(x)$, so $P'(h)(x) = g^x(P(f(h))(x)) = g^x(P(f(h'))(x)) = P'(h')(x)$. This establishes that P' respects our visibility.

Now, take any $h \in \prod_{x \in X} Q^x_-$. We will show that the set of incorrect guesses under the coloring h and the predictor P' is contained in the set of incorrect guesses under the coloring $f(h)$ and the predictor P, a set which we know must be in $\mathcal{I}$. Suppose x guesses incorrectly under the coloring h and the predictor P'; that is, $\neg h(x)Q^x(P'(h)(x))$. Then $\neg h(x)Q^x(g^x(P(f(h))(x)))$. By the definition $(f^x, g^x) : Q^x \to R^x$ being a morphism, it follows that $\neg f^x(h(x))R^x(P(f(h))(x))$; noting that $f^x(h(x)) = f(h)(x)$, this means that x guesses incorrectly under the coloring $f(h)$ and the predictor P. □

9.3   Two-Agent Problems and Morphisms

In our present context, two-agent problems are an important special case. Given relations Q, R, we use the pair (Q, R) to denote the two-agent (relational) hat problem in which agent 0 is assigned Q, agent 1 is assigned R, the agents see each other, and we require at least one correct guess. If (Q, R) admits a winning predictor, we call (Q, R) a *winning pair*.

Example 9.3.1. Given functions f, g, write $f \sim g$ if f and g agree infinitely often. Let $R = ({}^{\omega}\omega, {}^{\omega}\omega, \sim)$. Then the two-agent problem (R, R) is essentially the parity relation with ω colors: we think of the evens, collectively, as forming agent 0, while the odds, collectively, form agent 1. Technically, this captures the variant of the parity relation in which every even sees *every* odd and vice versa. We can come closer to capturing the parity relation as presented earlier by using $R = ({}^{\omega}\omega/{=}^{*}, {}^{\omega}\omega/{=}^{*}, \sim)$, where ${}^{\omega}\omega/{=}^{*}$ is the set of equivalence classes of ${}^{\omega}\omega$ under ${=}^{*}$; note that $\sim$ is still well defined here.

We now exhibit the connection between winning predictors in two-agent hat problems and morphisms between relations.

Theorem 9.3.2. (A, B) *is a winning pair iff* $A \preceq B^{\perp}$ *iff* $B \preceq A^{\perp}$. *More precisely,* $P = (S^0, S^1)$ *is a winning predictor for* (A, B) *iff* $(S^1, S^0) : A \to B^{\perp}$ *iff* $(S^0, S^1) :$ $B \to A^{\perp}$.

So, our results about minimal predictors for the parity relation can be seen as results about morphisms between various relations.

Proof. Given $S^0 : B_- \to A_+$ and $S^1 : A_- \to B_+$,

$$(S^0, S^1) \text{ is a winning predictor for } (A, B)$$
$$\iff \forall a \in A_- \ \forall b \in B_- \ (a A S^0(b) \lor b B S^1(a))$$
$$\iff \forall a \in A_- \ \forall b \in B_- \ (\neg b B S^1(a) \to a A S^0(b))$$
$$\iff \forall a \in A_- \ \forall b \in B_- \ (S^1(a) B^{\perp} b \to a A S^0(b))$$
$$\iff (S^1, S^0) : A \to B^{\perp}.$$

This establishes the first equivalence. The equivalence $(S^1, S^0) : A \to B^{\perp}$ iff $(S^0, S^1) : B \to A^{\perp}$ follows from the previous proposition. $\square$

To help visualize what is happening above, note that the diagrams that you might draw for a predictor (S^0, S^1) for (A, B) and a morphism $(S^1, S^0) : A \to B^{\perp}$ are the same except for a half-twist:

$$A_+ \quad {}_{S^0} \qquad {}_{S^1} \quad B_+ \qquad A_+ \xleftarrow{\;S^0\;} B_-$$

$$A_- \qquad\qquad\qquad B_- \qquad A_- \xrightarrow[\;S^1\;]{} B_+$$

Corollary 9.3.3. $A \preceq B$ iff $(A, B^\perp)$ is a winning pair.

Corollary 9.3.4. $(A, A^\perp)$ is always a winning pair.

Corollary 9.3.5. (A, A) is a winning pair iff $A \preceq A^\perp$.

Proposition 9.3.6.

(a) If A is not serial, then $B \preceq A$ for any relation B (provided $B_+ \neq \emptyset$).
(b) If $A^\perp$ is not serial, then $A \preceq B$ for any relation B (provided $B_- \neq \emptyset$).

Proof. For (a), fix $k \in A_-$ such that $\neg k A y$ for all $y \in A_+$; then $(x \mapsto k, g)$ is a morphism from A to B for any $g : A_+ \to B_+$. Dualizing yields (b). □

Theorem 9.3.7. If $A \circ A \circ A \subseteq A$ (in particular, if A is transitive), then the following are equivalent.

1. $A^\perp$ is serial.
2. $A^\perp \preceq A$.

Proof. Though we must have $A_- = A_+$ to form $A \circ A$, we will nevertheless write one or the other depending on which fits better logically.

For (1)⇒(2), suppose $A^\perp$ is serial. Interestingly, we apparently must break into cases depending on whether or not A is serial.

Case 1: A is serial (so A is decent). Then there exist functions $f : A_- \to A_+$ and $g : A_+ \to A_-$ such that $f \subseteq A$ and $g \subseteq A^\perp$. We claim that $(f, g) : A^\perp \to A$. Take any $x \in A_+ = A_-^\perp$ and $y \in A_+$, and suppose for a contradiction that $f(x) A y$ but $\neg x A^\perp g(y)$. Then $g(y) A x$. Noting that $x A f(x)$ (since $f \subseteq A$) and $f(x) A y$, we have $g(y)(A \circ A \circ A)y$, so $g(y) A y$ by our assumption on A. Then $\neg y A^\perp g(y)$, contradicting $g \subseteq A^\perp$.
Case 2: A is not serial. Then $A^\perp \preceq A$ is immediate from an earlier proposition. (A minor point: if A is not serial, then $A_- \neq \emptyset$, so $(A^\perp)_+ \neq \emptyset$.)

For (2)⇒(1), suppose $A^\perp$ is not serial. Fix $x \in A_+$ such that $y A x$ for all y. Then for any proposed morphism (f, g), we have $f(x) A x$, but $\neg x A^\perp f(x)$, so (f, g) is not a morphism. □

Corollary 9.3.8. If A is decent and transitive, then $(A^\perp, A^\perp)$ is a winning pair.

9.4 Norms

Definition 9.4.1. Let A be a relation. A *dominating set for* A is a $D \subseteq A_+$ such that $\forall x \in A_- \exists y \in D \; xAy$, and an *indominable set for* A is an $I \subseteq A_-$ such that $\forall y \in A_+ \exists x \in I \; \neg xAy$; note that the indominable sets for A are the dominating sets for $A^\perp$. The *norm* $\|A\|$ of A is the least cardinality of a dominating set for A, so $\|A^\perp\|$ is of course the least cardinality of an indominable set for A. If A has no dominating set, we write $\|A\| = \infty$.

Note that A is decent iff A_+ is a dominating set for A and A_- is an indominable set for A. So, when A is decent, we have $\|A\| \leq |A_+|$, $\|A^\perp\| \leq |A_-|$.

Lemma 9.4.2. *If* $A \preceq B$, *then* $\|A\| \leq \|B\|$.

Proof. Given $(f, g) : A \to B$, if D is a dominating set for B, then $g[D]$ is a dominating set for A. □

The implication (4) $\Rightarrow$ (1) in the proof of Theorem 4.5.6 is essentially a special case of the following.

Corollary 9.4.3. *If* $\|A\| > \|B^\perp\|$, *then* (A, B) *is not a winning pair.*

Proof. Apply Theorem 9.3.2 and the above lemma. □

9.5 Applications of the Metaphor

An immediate payoff of viewing morphisms in terms of hat problems is that the different point of view sometimes lends itself to intuitive proofs. We offer two examples.

Theorem 9.5.1. *If* A *is transitive, then* $\|A\| \leq \|A^\perp\|$ *iff* $A \preceq A^\perp$.

Proof. Lemma 9.4.2 gives us the right-to-left direction immediately.

Suppose A is transitive and $\|A\| \leq \|A^\perp\|$. We establish $A \preceq A^\perp$ by showing that (A, A) is a winning pair. Let $\kappa = \|A\|$ and let $D = \{d_\alpha : \alpha < \kappa\}$ be a dominating set for A (with $d_\alpha \neq d_\beta$ for $\alpha \neq \beta$). Define $r : A_- \to \kappa$ by letting $r(x)$ be minimal such that $xAd_{r(x)}$. Informally, our predictor works as follows: under coloring (x^0, x^1), agent i produces a guess that, under the assumption $r(x^i) \leq r(x^{1-i})$, would be correct; this will be a minimal predictor because at least one of these two assumptions must be true. Specifically, define $f : A_- \to A_+$ such that for all $x \in A_-$, we have $d_\alpha Af(x)$ for all $\alpha \leq r(x)$; we can do this because $r(x) < \|A\| \leq \|A^\perp\|$, so the set $\{d_\alpha : \alpha \leq r(x)\}$ is not indominable. Our predictor is $S = (f, f)$. Given any coloring (x^0, x^1), let i be such that agent i's assumption $r(x^i) \leq r(x^{1-i})$ is correct. Then $x^i Ad_{r(x^i)} Af(x^{1-i})$, so $x^i Af(x^{1-i})$ by the transitivity of A, making agent i's guess correct. Therefore, (A, A) is a winning pair, so $A \preceq A^\perp$. □

Our second theorem is a result from [Yip94] that was considered somewhat surprising when it first was established.

Theorem 9.5.2. *If* $\|A\| = |A_+| = \|B^\perp\| = |B_-|$, *then there is a morphism from* B *to* A.

Proof. Let $\kappa = \|A\| = |A_+| = \|B^\perp\| = |B_-|$ and fix bijections $\rho : A_+ \to \kappa$, $\rho' : B_- \to \kappa$.

For any $a \in A_+$, define $D_a = \{b \in B_- : \rho'(b) \leq \rho(a)\}$. Noting that $|D_a| < \kappa = \|B^\perp\|$, we can let $g(a) \in B_+$ such that $bBg(a)$ for all $b \in D_a$. Similarly, for $b \in B_-$, define $E_b = \{a \in A_+ : \rho(a) \leq \rho'(b)\}$; $|E_b| < \kappa = \|A\|$, so we can let $f(b) \in A_-$ such that $\neg f(b)Aa$ for all $a \in E_b$.

Take any $b \in B_-$, $a \in A_+$; if $f(b)Aa$, then $a \notin E_b$, so $\rho'(b) < \rho(a)$, so $b \in D_a$, so $bBg(a)$. Therefore, $(f, g) : B \to A$. □

A potential but as yet unrealized payoff is that the present point of view offers a way in which morphisms are a special case of a more general notion (predictors), which invites us to take known facts (or questions) about morphisms that have proven to be of value and consider their analogs in the more general context of predictors. For example, in terms of hat problems, the Cichoń diagram gives us a certain set of winning pairs (for instance, $(\mathbf{R}, \mathcal{M}, \in) \preceq (\mathcal{N}, \mathbf{R}, \not\ni)$ gives us that $((\mathbf{R}, \mathcal{M}, \in), (\mathbf{R}, \mathcal{N}, \in))$ is a winning pair). We can consider the ternary analog of this: with agents 0, 1, 2 and visibility forming a cycle, we can call a triple (R^0, R^1, R^2) of acceptability relations a *winning triple* if it admits a minimal predictor; the set of winning triples where the R^i are drawn from $(\mathbf{R}, \mathcal{M}, \in)$, $(\mathcal{M}, \mathcal{M}, \subseteq)$, $(\mathbf{R}, \mathcal{N}, \in)$, $(\mathcal{N}, \mathcal{N}, \subseteq)$ and their duals is then a ternary analog of the Cichoń diagram. We get another analog if we let each of the three agents see both other agents. Perhaps one or both of these analogs is of interest?

9.6 Scattered-Error Predictors

Here, we show that there always exist decent notions of acceptability that are sufficiently easy to make a scattered-error predictor exist.

For a proximity scheme V on X, say that $\Sigma \subseteq X$ is *scattered* if every nonempty $\Sigma' \subseteq \Sigma$ has an x with proximity V such that $V \cap \Sigma' = \emptyset$. When V is induced by a topology by letting V_x be the filter of sets containing a deleted neighborhood of x, this coincides with the existing definition of scattered.

Theorem 9.6.1. *For every proximity scheme, there exists a decent acceptability notion that admits a scattered-error predictor.*

In contrast to results based on the μ-predictor, we can prove this in ZF, and the predictor is constructed quite directly.

Proof. Let λ be a limit ordinal such that every x has a proximity whose cardinality is strictly less than $cf(\lambda)$. (For infinite X, $\lambda = |X|^+$ always works. For finite X, $\lambda = \omega$ works.) Let $A = (\lambda, \lambda, \leq)$. Define the predictor P by letting $P(f)(x)$ be any α such that x has a proximity U with $\alpha \geq f(y)$ for all $y \in U$. (For instance, we could take the least such α.) We can verify that this respects $\approx_x$.

Let f be any coloring, and let W be the resulting set of wrong guesses. We claim that W is scattered. Take any nonempty $W' \subseteq W$. Let $x \in W'$ be such that $P(f)(x)$ is minimal among $\{ P(f)(y) : y \in W' \}$. Let U be a proximity of x such that $P(f)(x) \geq f(y)$ for all $y \in U$. We claim that U isolates x from W'. If $y \in W' \cap U$, then $P(f)(y) \geq P(f)(x) \geq f(y)$, so y guesses correctly, a contradiction. Therefore $W' \cap U = \emptyset$, establishing that W is scattered. $\square$

Say that a visibility graph V is *thin* if it contains no infinite chain, and *thick* if its complement is thin. While we still do not know whether the thick visibility graphs are exactly those that admit finite-error predictors for any set of colors, we can at least conclude the following.

Corollary 9.6.2. *Let V be a visibility graph on X. Then the following are equivalent.*

1. V is thick.
2. V admits a finite-error predictor for some decent notion of acceptability.

Proof. It is straightforward to show that V is thick iff every scattered set is finite.

                $\square$

Corollary 9.6.3. *V has an infinite path (or a cycle) iff there is a decent A such that, assigning A to each agent, there is a minimal predictor.*

Proof. It is straightforward to show that V has an infinite path or a cycle iff the set of all agents is not scattered. $\square$

Corollary 9.6.4. *In any topological space, not necessarily T_0, there is a decent A that yields a scattered-error predictor.*

Proof. Trivial. $\square$

Corollary 9.6.5. *With complete visibility, there is a decent A that yields a predictor guaranteeing at most one error.*

Proof. With complete visibility, scattered sets have at most one point. $\square$

This special case is obvious when you consider what the predictor is doing: guess something at least as big as everything you see. The only way to fail is if your color is strictly greater than everyone else's, and that can only happen to one player.

9.7   Pseudo-scattered Sets

Let V be a proximity scheme on X. We know there exist decent relations that admit scattered-error predictors. So now we can turn to the question, Which ones? Or at least, what sufficient condition can we find? We do not have satisfying answers to either question, but we can at least make some headway if we switch to *pseudo-scattered* sets, which often coincide with scattered sets.

Definition 9.7.1. Given a relation V on X, a set $\Sigma \subseteq X$ is *pseudo-scattered* if it contains no infinite V-chain.

Proposition 9.7.2. *The pseudo-scattered subsets of X form an ideal (and this ideal contains all singletons).*

Proposition 9.7.3. *If V is transitive, the pseudo-scattered sets are exactly the co-well-founded sets (which are the scattered sets when one takes the upward topology).*

Definition 9.7.4. Say that the *transitivity width* of a binary relation V is the smallest size of a family $\mathcal{T}$ of transitive subrelations of V such that every (infinite) chain $C \subseteq V$ contains an infinite C' such that C' is a chain in some element of $\mathcal{T}$.

Example 9.7.5. If V is transitive, we can use $\mathcal{T} = \{V\}$, making the transitivity width ≤ 1. Similarly, if V has a transitive subrelation V' such that $V - V'$ is thin, we can use $\mathcal{T} = \{V'\}$.

If V is thin, then we can use $\mathcal{T} = \emptyset$, making the transitivity width of V 0. (Note that the transitivity width is 0 iff V is thin.)

Theorem 9.7.6. *Let V be a visibility graph on X. Suppose $\|A^\perp\|$ is strictly greater than the transitivity width of V. Then, using A as the acceptability relation for each agent, there exists a pseudo-scattered-error predictor.*

As a slight refinement, one can see in the following proof that there is no need to assign the same acceptability relation to each agent; we just need each acceptability relation A to have $\|A^\perp\|$ strictly greater than the transitivity width.

Proof. Let $\mathcal{T}$ be a family as in Definition 9.7.4, with

$$\|A^\perp\| > |\mathcal{T}|. \tag{9.1}$$

For each $W \in \mathcal{T}$, let M_W be the corresponding μ-predictor, and for any coloring f, let $\langle f \rangle_x^W$ be like $\langle f \rangle_x$ but with visibility given by W instead of V. We form the predictor P by choosing $P(f)(x) \in A_+$ such that for every $W \in \mathcal{T}$, $(M_W(f)(x))A(P(f)(x))$; we can do so by (9.1).

Suppose the set of errors is not pseudo-scattered for some coloring f. Let C be an infinite chain of wrong guessers in V. Let $C' \subseteq C$ be infinite such that C' is a chain in some $V' \in \mathcal{T}$. Chose $x \in C'$ such that $\langle f \rangle_x^{V'}$ is minimal among $\{ \langle f \rangle_y^{V'} : y \in C' \}$. Let $y \in C'$ be such that $x V' y$. Then $\langle f \rangle_y^{V'} \preceq \langle f \rangle_x^{V'}$ (since, under V', y sees a subset of what x sees), so $\langle f \rangle_y^{V'} = \langle f \rangle_x^{V'}$ by the minimality of the latter, so $M_{V'} f(y) = \langle f \rangle_y^{V'}(y) = \langle f \rangle_x^{V'}(y) = f(y)$. We have $(M_{V'} f(y))A(P(f)(y))$, so $f(y)A(P(f)(y))$, so y guessed correctly, a contradiction. $\square$

Here is what goes wrong if one tries to show that P is a scattered-error predictor: if Σ is a set of wrong guesses that is not scattered, it might still be the case that Σ contains no infinite chain of V, and it would be that infinite chain that we would use to get a contradiction.

Bibliography

[ABFR94] Aspnes, J., Beigel, R., Furst, M., Rudich, S.: The expressive power of voting polynomials. Combinatorica **14**, 135–148 (1994)

[BM90] Banaschewski, B., Moore, G.H.: The dual Cantor-Bernstein theorem and the partition principle. Notre Dame J. Form. Logic **31**(3), 375–381 (1990)

[Bar10] Bartoszyński, T.: Invariants of measure and category. In: Foreman, M., Kanamori, A. (eds.) Handbook of Set Theory, vol. 1, pp. 491–555. Springer, Dordrecht (2010)

[BJ95] Bartoszyński, T., Judah, H.: Set theory: on the Structure of the Real Line. A K Peters, Wellesley (1995)

[Bla10] Blass, A.: Combinatorial cardinal characteristics of the continuum. In: Foreman, M., Kanamori, A. (eds.) Handbook of Set Theory, vol. 1, pp. 395–489. Springer, Dordrecht (2010)

[BT78] Baumgartner, J.E., Taylor, A.D.: Partition theorems and ultrafilters. Trans. Am. Math. Soc. **241**, 283–309 (1978)

[BTW82] Baumgartner, J.E., Taylor, A.D., Wagon, S.: Structural properties of ideals. Dissertationes Math. **197**, 1–99 (1982)

[B58] Berge, C.: Sur le couplage maximum d'un graphe. Comptes rendus hebdomadaires des séances de l'Académie des sciences **247**, 258–259 (1958)

[Bla94] Blass, A.: Cardinal characteristics and the product of countably many infinite cyclic groups. J. Algebra **169**, 512–540 (1994)

[Bla96] Blass, A.: Reductions betweencardinal characteristics ofthe continuum. In: Bartoszyński, T., Scheepers, M. (eds.) Contemporary Mathematics 192: Set Theory: Annual Boise Extravaganza in Set Theory (BEST) Conference, pp. 31–49 (1996)

[Bre95] Brendle, J.: Evasion and prediction – the Specker phenomenon and Gross spaces. Forum Math. **7**, 513–541 (1995)

[Bre03] Brendle, J.: Evasion and prediction III: constant prediction and dominating reals. J. Math. Soc. Jpn. **55**(1), 101–115 (2003)

[BS96] Brendle, J., Shelah, S.: Evasion and prediction II. J. Lond. Math. Soc. **53**, 19–27 (1996)

[BS03] Brendle, J., Shelah, S.: Evasion and prediction IV: strong forms of constant prediction. Arch. Math. Logic **42**, 349–360 (2003)

[Buh02] Buhler, J.P.: Hat tricks. Math. Intell. **24**(4), 44–49 (2002)

[BHKL08] Butler, S., Hajiaghayi, M.T., Kleinberg, R.D., Leighton, T.: Hat guessing games. SIAM J. Discret. Math. **22**, 592–605 (2008)

[Can66] Cantor, G.: Gesammelte Abhandlungen Mathematischen und Philosophischen Inhalts. Georg Olms, Hildesheim (1966)

C.S. Hardin and A.D. Taylor, *The Mathematics of Coordinated Inference*,
Developments in Mathematics 33, DOI 10.1007/978-3-319-01333-6,
© Springer International Publishing Switzerland 2013

[CS84] Carlson, T.J., Simpson, S.G.: A dual form of Ramsey's theorem. Adv. Math. **53**, 265–290 (1984)

[DG76] Daviesm, R.O., Galvin, F.: Solution to query 5. Real Anal. Exch. **2**, 74–75 (1976)

[Ebe98] Ebert, T.: Applications of recursive operators to randomness and complexity. PhD thesis, University of California at Santa Barbara (1998)

[E83] Eda, K.: On a Boolean power of a torsion-free abelian group. J. Algebra **82**, 84–93 (1983)

[EH66] Erdős, P., Hajnal, A.: On a problem of B. Jonsson. Bull. Acad. Polon. Sci. Ser. Math. Astron. Phys. **14**, 19–23 (1966)

[EHMR84] Erdős, P., Hajnal, A., Máté, A., Rado, R.: Combinatorial Set Theory: Partition Relations for Cardinals. North-Holland, Amsterdam (1984)

[FHMV95] Fagin, R., Halpern, J.Y., Moses, Y., Vardi, M.Y.: Reasoning About Knowledge. MIT, Cambridge (1995)

[Fei04] Feige, U.: You can leave your hat on (if you guess the color). Technical report MCS04-03, Computer Science and Applied Mathematics, The Weizmann Institute of Science, p. 10 (2004)

[Fre90] Freiling, C.: Symmetric derivates, scattered, and semi-scattered sets. Trans. Am. Math. Soc. **318**, 705–720 (1990)

[Gal65] Galvin, F.: Problem 5348. Am. Math. Mon. **72**, 1136 (1965)

[GP76] Galvin, F., Prikry, K.: Infinitary Jonsson algebras and partition relations. Algebra Univ. **6**(3), 367–376 (1976)

[Gar61] Gardner, M.: The 2nd Scientific American Book of Mathematical Puzzles & Diversions. Simon and Schuster, New York (1961)

[Geo07] George, A.: A proof of induction? Philos. Impr. **7**(2), 1–5 (2007)

[Gil02] Gillman, L.: Two classical surprises concerning the axiom of choice and the continuum hypothesis. Am. Math. Mon. **109**(6), 544–553 (2002)

[GR71] Graham, R., Rothschild, B.: A survey of finite Ramsey theory. In: Proceedings of the 2nd Louisiana Conference on Combinatorics, Graph Theory and Computing, Baton Rouge, pp. 21–40 (1971)

[G71] Grigorieff, S.: Combinatorics on ideals and forcing. Ann. Math. Logic **3**(4), 363–394 (1971)

[H10] Hardin, C.S.: On transitive subrelations of binary relations. J. Symb. Logic **76**, 1429–1440 (2010)

[H13] Hardin, C.S.: Universality of the μ-predictor. Fundam. Math. **208**, 227–241 (2013)

[HT08a] Hardin, C.S., Taylor, A.D.: An introduction to infinite hat problems. Math. Intell. **30**(4), 20–25 (2008)

[HT08b] Hardin, C.S., Taylor, A.D.: A peculiar connection between the axiom of choice and predicting the future. Am. Math. Mon. **115**(2), 91–96 (2008)

[HT09] Hardin, C.S., Taylor, A.D.: Limit-like predictability for discontinuous functions. Proc. AMS **137**, 3123–3128 (2009)

[HT10] Hardin, C.S., Taylor, A.D.: Minimal predictors in hat problems. Fundam. Math. **208**, 273–285 (2010)

[Hen59] Henkin, L.: Some remarks on infinitely long formulas. In: Infinitistic Methods, Proceedings of the Symposium on Foundations of Mathematics, Warsaw, pp. 167–183 (1959)

[HW13] Horsten, L., Welch, P.: The aftermath. Math. Intell. **35**(1), 16–20 (2013)

[Hyt06] Hyttinen, T.: Cardinal invariants and eventually different functions. Bull. Lond. Math. Soc. **38**, 34–42 (2006)

[Jec03] Jech, T.J.: Set Theory: The Third Millenium Edition, Revised and Expanded. Springer Monographs in Mathematics. Springer, Berlin (2003)

[JS93] Judah, H., Shelah, S.: Baire property and axiom of choice. Isr. J. Math. **84**, 435–450 (1993)

[Kad98] Kada, M.: The Baire category theorem and the evasion number. Proc. Am. Math. Soc. **126**(11), 3381–3383 (1998)

[Kam00] Kamo, S.: Cardinal invariants associated with predictors. In: Buss, S. et al. (eds.) Logic Colloquium'98, Prague. Lecture Notes in Logic, vol. 13, pp. 280–295 (2000)

[Kam01] Kamo, S.: Cardinal invariants associated with predictors II. J. Math. Soc. Jpn. **53**, 35–57 (2001)

[Kun71] Kunen, K.: Elementary embeddings and infinitary combinatorics. J. Symb. Logic **36**, 407–413 (1971)

[Kun80] Kunen, K.: Set Theory. Elsevier, Amsterdam (1980)

[LS02] Lenstra, H., Seroussi, G.: On hats and other covers. In: IEEE International Symposium on Informations Theory, Lausanne (2002)

[Mal66] Malitz, J.: Problems in the model theory of infinite languages. PhD thesis, University of California, Berkeley (1966)

[Mat77] Mathias, A.R.D.: Happy families. Ann. Math. Logic **12**, 59–111 (1977)

[Mil81] Miller, A.W.: Some properties of measure and category. Trans. Am. Math. Soc. **266**(1), 93–114 (1981)

[Mild97] Mildenberger, H.: Non-constructive Galois-Tukey connections. J. Symb. Logic **62**(4), 1179–1186 (1997)

[Pas08] Paseau, A.: Justifying induction mathematically: strategies and functions. Log. Anal. **203**, 263–269 (2008)

[Mor90] Morgan II, J.: Point Set Theory. Marcel Dekker, Inc., New York (1990)

[Ram30] Ramsey, F.: On a problem of formal logic. Proc. Lond. Math. Soc. **30**, 264–286 (1930)

[She84] Shelah, S.: Can you take Solovay's inaccessible away? Isr. J. Math. **48**, 1–47 (1984)

[SS00] Shelah, S., Stanley, L.: Filters, Cohen sets and consistent extensions of the Erdős-Dushnik-Miller theorem. J. Symb. Logic **65**, 259–271 (2000)

[Sie47] Sierpiński, W.: L'hypothèse généralisée du continu et l'axiome du choix. Fundam. Math. **34**, 1–5 (1947)

[Sil66] Silverman, D.L.: Solution of problem 5348. Am. Math. Mon. **73**, 1131–1132 (1966)

[Sol70] Solovay, R.: A model of set theory in which every set of reals is Lebesgue measurable. Ann. Math. **92**, 1–56 (1970)

[S50] Specker, E.: Additive Gruppen von Folgen ganzer Zahler. Port. Math. **9**, 131–140 (1950)

[T12] Taylor, A.D.: Prediction problems and ultrafilters on ω. Fundam. Math. **219**, 111–117 (2012)

[Tho67] Thorp, B.L.D.: Solution of problem 5348. Am. Math. Mon. **74**, 730–731 (1967)

[Tu47] Tutte, W.T.: The factorization of linear graphs. J. Lond. Math. Soc. **22**, 107–111 (1947)

[Vel11] Velleman, D.J.: The even-odd hat problem (2011, preprint)

[V93] Vojtáš, P.: GeneralizedGalois-Tukey connections between explicit relations on classical objects of real analysis. In: Judah, H. (ed.) Set Theory of the Reals, Israel Mathematical Conference Proceedings, vol. 6, pp. 619–643. American Mathematical Society, Providence (1993)

[Wim82] Wimmers, E.: The Shelah P-point independence theorem. Isr. J. Math. **4**, 28–48 (1982)

[Win01] Winkler, P.: Games people don't play. In: Wolfe, D., Rodgers, T. (eds.) Puzzlers' Tribute, pp. 301–313. A K Peters, Natick (2001)

[Yip94] Yiparaki, O.: On some tree partitions. PhD thesis, University of Michigan (1994)

Index

C.S. Hardin and A.D. Taylor, *The Mathematics of Coordinated Inference*,
Developments in Mathematics 33, DOI 10.1007/978-3-319-01333-6,
© Springer International Publishing Switzerland 2013

Printed in the United States
By Bookmasters